ENCYCLOPAEDIA OF ENDOCRINOLOGY - V

MAMMALIAN ENDOCRINOLOGY

By

Manju Yadav

Lecturer

Department of Zoology

M.M.H. College

Ghaziabad (U.P.)

(India)

DISCOVERY

PUBLISHING HOUSE

First Published – 2008

Reprinted – 2025

ISBN: 978-93-5056-570-4 (Set)
978-81-8356-275-1

Mammalian Endocrinology

Published by:

DISCOVERY PUBLISHING HOUSE

4383/4B, Ansari Road, Darya Ganj
New Delhi-110 002 (India)
Phone: +91-11-23279245; 23253475; 43596065
Mobile: +91 9811179893 / +91 9871656464
E-mail: discoverybooksindia@gmail.com
orderdphbooks@gmail.com
namitwasan9@gmail.com
web: www.discoverypublishinggroup.com

Printed at:
Infinity Imaging Systems
Delhi

Preface

The present title **Mammalian Endocrinology** provides a definitive and comprehensive review of all aspects relating to hormones in different animals including wild and domestic species. It discusses the intimate physiology of the endocrine system itself and describes the role of hormones in the processes of nutrition, osmoregulation, colour change, calcium metabolism and reproduction. Subject matter has been presented in a readable way emphasizing the differences between species from a functional aspects. It not only contains a wealth of information concerning the endocrinology of the various species but also contains more than enough basic information concerning the biosynthesis, metabolism and mode of action of the hormones for the book to be read and understood without referring to a text book of basic endocrinology. This book has been written primarily for use as a text book by under-graduate, as well as graduate students.

Efforts have also been made to make the presentation lucid and accessible for beginning students, emphasizing major concepts while in corporating the latest experimental findings and theories throughout. The aim is to enthuse the readers with this active and exciting area of research and to lay a solid foundation on which further study of its various facts may be based.

Though, the author has taken special care to present a current account, yet he is fully aware about limitations and the readers may come across the mistakes of various types. For all types of mistakes author extends his due apology.

There can be no claim to originality except in the manner of treatment and much of the information has been obtained from the books and scientific journals available in the different libraries.

The author expresses his thanks to his friends and colleagues whose continue inspirations have initiated him to bring out this book.

The author expresses his gratitude to Mr. Wasan and staff of M/s Discovery Publishing House for their whole hearted co-operation in the publication of this book.

Author

CONTENTS

1

INTRODUCTION

In a previous summary evidence was outlined showing that in female rats different pituitary regulatory mechanisms, neural or otherwise, are responsible: (1) for follicle growth and estrogen secretion; (2) for preovulatory maturation of the follicles, ovulation and luteinization *per se*; and (3) for activation of organized corpora lutea. It seemed clear then that although much was known about the ovulation phase (2) the significance of neural mechanisms in (1) and (3) had yet to be defined.

The ovulation phase of the cycle is best understood for the simple reasons that, on the one hand, this phase is most accessible to experimental attack, especially in rabbits and rats, and, on the other hand, it is a climactic event of primary importance and interest. As the literature dealing with reproductive physiology is examined, it appears that a large proportion of investigations ostensibly addressed to the general question of control of gonadotropin secretion are in actual fact concerned with the ovulation phase alone.

Taking the rat to be characteristic of spontaneously ovulating mammals, and the rabbit to be characteristic of reflexly ovulating ones, the following statements may legitimately be made about this phase of the generalized mammalian cycle: (1) Reflex ovulation and spontaneous ovulation, alike, are governed by a hypothalamico-pituitary apparatus whose final link to the pars distalis is neurohumoral, via the hypophyseal portal venules, and whose activity precipitates the release of ovulating hormone. (2) The apparatus includes a hypothalamic center or constellation of centers whose excitation, or inhibition, depends on circulating estrogen and progestogen, and on afferent nerve impulses of varying kind. (3) Thresholds in the several components of the apparatus are influenced not only by the sex steroids but by other factors, poorly understood indeed, that vary from animal to animal and from time to time in a given individual, e.g., the diurnal rhythm in rats. (4) Dual mechanisms subserving reflex ovulation and spontaneous ovulation are potentially present in some, if not all, species.

We cannot now go far beyond these statements in describing either the mechanism or its essential components and their respective roles. This field is still developing slowly. For example, Hohlweg and Junkmann demonstrated in 1932 that estrogen will induce corpora lutea in prepuberal rats, but more than 10 years passed before this was found to be true in adult mammals. Even now we know it to be true in only a pitifully few species. Facilitation of ovulation by progesterone similarly has been shown in only three mammals, although the first observation was reported in 1940. The view that the ovulation-inducing actions of these steroids is mediated by the hypothalamus in rats and rabbits seems uncontested, but their sites of action remain obscured.

To be sure, we know a little more than we did seven years ago. We know, for example, several additional substances that will block spontaneous ovulation in rats. These include morphine, reserpine, chlorpromazine, Pathilon (*3-diethylammo-1-cyclohexyl-1-phenyl-1-ethiodide*) and SKF-1045A (*6-methyl-2-aminobenzothiazole*). It is now established that the "critical period" during which rats spontaneously stimulate their pituitaries occurs between 2:00 and 4:00 P.M. in commercial rats housed briefly in Southern California, just as predictably as in our inbred rats in North Carolina. The common factor is the controlled daily illumination of 14 hours.

In further analysis of that "critical period," it has been established that release of an adequate amount of ovulating hormone from the rat pituitary takes roughly half an hour. Furthermore, this process goes on *pari passu* with the atropine-sensitive component of the "LH-release mechanism". Hence, in rats, this mechanism is less like a trigger than it is in rabbits.

An observation by Critchlow (1957) is especially important, on two counts. Working with proestrous rats injected with a blocking dose of nembutal shortly before the "critical period," he placed bipolar stimulating electrodes variously in different hypothalamic regions. When the electrode tips were deep in the tuber just above the median eminence, ovulation was frequently induced by stimulations extending over approximately 30 minutes. The nembutal blockade was thus by-passed.

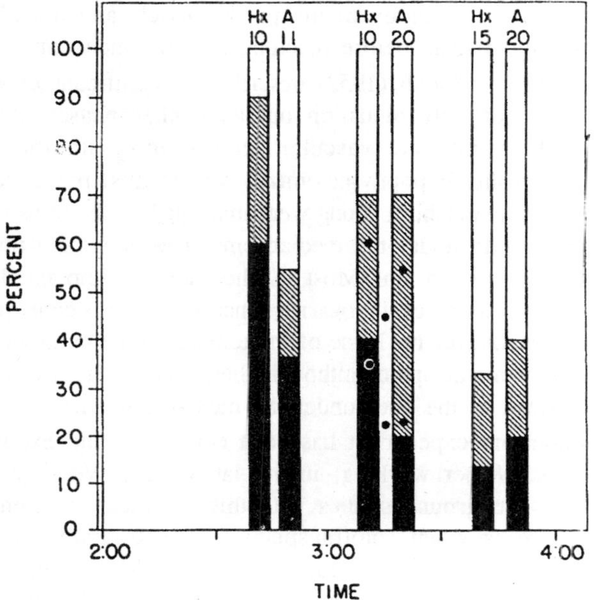

Fig. 1.1. Analysis of the "critical period" in a homogeneous strain of rats by means of hypophysectomy (Hx) or atropine injection (A) at different times.

Similarly placed electrodes without stimulation were ineffective, as were stimuli delivered through electrodes further removed from the median eminence. This finding is important not only because it emphasizes the importance of the stalk mechanism in this spontaneously ovulating species, but also because it constitutes the first published account of ovulation induced in a spontaneously ovulating mammal by stimulation of the brain.

The second such report was published recently by Bunn and Everett (1957). In this instance, we blocked ovulation in rats by a regime of continuous illumination and subsequently induced ovulation by stimulation of the amygdaloid nucleus. Thus one can now say that electrical stimulation of the amygdala will somehow invoke ovulation, not only in rabbits and cats, but also in this species which normally ovulates spontaneously.

Although evidence strongly favours the view that the final link to the pars distalis is a chemotransmitter delivered to the gland by the hypophyseal portal vessels, the identity of the substance remains to be established. Perhaps it is an adrenergic substance, as postulated, and is subject to blockade at the pars distalis by Dibenamine congeners. That, however, is still to be proved, and we need to know much more about the special physiology of the portal vessels and the glandular cells which they irrigate before reaching a firm conclusion.

Harris and Jacobsohn (1952) described a significant experiment indicating that not only ovulation but also other phases of the rat cycle depend on intimate vascular relationship with the median eminence. In freshly hypophysectomized female rats, pituitaries from the animals' own newly born young were implanted by the trans-temporal route either directly under the median eminence or simply under the temporal lobe of the brain. Most of the mothers implanted in the former manner regained cyclic ovarian function; many became pregnant and delivered viable litters. None of the temporal lobe controls cycled. Their ovaries were atrophic, although their traits were as large and well vascularized as the ones under the median eminence.

This important experiment has been confirmed and extended by Dr. Nikitovitch-Winer working; in my laboratory. We may quickly pass over the background studies with this commentary: Female rat anterior pituitaries when autotransplanted to extracranial sites will selectively maintain functional corpora lutea while the remainder of the ovary atrophies. Luteotropin secretion is seemingly favoured by the new circumstances and continues uninterruptedly as long as desired. Secretion of FSH and LH, on the other hand, is promptly lost, at least as far as can be judged from ovarian histology.

Winer first transplanted the pars distalis of an adult, cycling rat to the renal capsule, allowed the graft to undergo the changes just mentioned for approximately one month, and then re-transplanted it back into close relationship with the median eminence. Control groups were three in number: (1) animals bearing re-transplants under the temporal lobe; (2) animals retaining autografts on the kidney; and (3) normal, cycling adults. None of the temporal lobe or kidney controls gave evidence of FSH or LH secretion, but in the median eminence group, 13 of 14 rats returned to cycling spontaneously. Onset of the first proestrus was 8 to 69 days after the second operation, the average being 36 days. All the cycling rats were mated with known fertile males: 7 accepted coitus and carried litters to term; 3 delivered spontaneously, 3 were delivered by caesarian section a day early, and the other on day 24.

The ovaries and reproductive tracts were well restored in cycling animals of the median eminence series, but were atrophic in the temporal lobe and kidney controls. The persistence of large functional corpora lutea was avoided in these experiments by carrying out the original transplantations during proestrus. The follicular apparatus and interstitial tissue were generally well restored in the median eminence group, although some localized portions of the ovaries remained inadequately stimulated. The average ovarian weight was well above the ovarian weights noted in the temporal lobe and kidney control groups, although it was still less than normal. This intermediate degree of stimulation was reflected also in a lower than normal number of follicles or corpora lutea in any given set and in the numbers of fetuses in the pregnant animals. This quantitative deficiency can easily be related to the small size of the graft compared to that of the normal gland. Similar intermediate degrees of restoration were noted also with respect to thyroid and adrenocortical stimulation, probably for the same reason. In view of the severe damage to the pars distalis caused by the two successive transplantations, it is surprising that even this much restoration was obtained.

Twenty-four hours after transplantation to the kidney or back to the median eminence, one finds that the major portion of glandular parenchyma is involved in a massive internal infarction and that healthy parenchyma remains in only a thin outer shell. Within a week, however, this healthy tissue becomes a well-organized, well-vascularized, compact organ and the infarct is reduced to a relatively minor connective tissue core. In kidney grafts the large basophiles, especially the large delta cells (gonadotrophs), disappear gradually during the first two or three weeks. By the end of a month only occasional minute ones persist among multitudes of small chromophobes and modified acidophiles. The same picture holds true for grafts re-transplanted under the temporal lobe. Grafts under the temporal lobe or in the kidney never show large gonadotrophs or castration (signet ring) cells. Grafts from the median eminence series contain them in abundance. The more fertile rats with regular cycles have few, if any, castration cells, and these are present in important numbers in only those animals that have irregular cycles or that remain anestrous.

Can there be any doubt that even the follicular phase of the cycle requires some special influence mediated by the median eminence? That we are dealing here with vascular rather than with nerve regeneration is strongly indicated by the studies of Harris (1950) and Harris and Jacobsohn (1950), and by the fact that in three of our median eminence experiments estrus returned within 8, 11, and 15 days, respectively. This is an inordinately rapid recovery to be attributable to nerve regeneration.

Space permits only a brief commentary on the luteal phase of the cycle. It is reasonable to assume that the pseudopregnancy cycle in rats and mice is essentially comparable to the generalized mammalian cycle. In the author's opinion, three aspects of the luteal phase require separate consideration: initiation, maintenance, and termination.

Ovulation and corpus luteum formation do not in themselves constitute the *initiation* of the functional luteal phase. This truly begins when progestogen secretion reaches such a rate that the uterus will be adequately prepared for nidation and forthcoming cycles will be suppressed. Corpus luteum formation is involved in preparation for this phase, certainly. Also, attainment of a state of hypothalamico-pituitary function which favours secretion of luteotropin (LTH) is required. This state may pre-exist, as it seems to in the case of the generalized mammalian cycle, or it may require some special stimulus to the hypothalamus, as in the rat. Under special circumstances it may be caused to "pre-exist" even in that animal (e.g., delayed pseudopregnancy). When elevated LTH secretion exists together with responsive corpora lutea, the rate of progesterone secretion mounts to high level and the luteal phase may then be said to have begun.

Maintenance of this high rate must require: (1) that LTH output remain high; (2) that no influence from other parts of the body or of the ovary, or from any factor intrinsic within the corpus luteum itself, lower its responsiveness to LTH. Our experiments with pituitary transplants on the kidney show that in rats, at least, no such intrinsic factor limits the life span of corpora lutea. Given a continuing supply of LTH they will continue to function indefinitely.

We should look elsewhere for mechanisms that *terminate* the luteal phase. After all, there is evidence not only from the rat, but from other species as well, that the corpus luteum life span is not built-in, as witness the well-known persistent corpus luteum of the cow.

A recent experiment in the mouse by Muhlbock and Boot (1956) is instructive with respect to this stage of the cycle. These workers noted that when an adult, cycling mouse is engrafted subcutaneously with a homotransplant of anterior pituitary, a sequence of pseudopregnancies supplants the short cycles. We have confirmed this in the rat, in its essentials. In such a preparation it is probable that a continuous supply of LTH comes from the graft, in quantity sufficient by itself to maintain a given set of corpora lutea indefinitely. Nevertheless, such animals periodically interrupt their pseudopregnancies, destroy the old corpora lutea, and form new ones. It is interesting to note the indication of a positive luteolytic effect associated with each recurring estrus and ovulation, in spite of the presumably unvarying secretion of LTH by the graft. One is reminded of an early experiment by Takewaki (1931) with parabiotic rats. When one female member of a pair of normal adult parabionts was made pregnant, the pregnancy would be interrupted if the partner was castrated during the progravid phase, presumably because of destruction of the corpora lutea in the former animal. In any event, in the Muhlbock experiment, we have clear evidence of a long cycle in the hypothalamico-pituitary mechanism. Although the short cycles so characteristic of rats and mice are suppressed by action of the extra LTH from the graft, the long cycle must break through eventually. It is the author's opinion, that here is the key to the overall periodicity of the generalized type of polyestrous, infertile mammalian cycles.

There is a striking analogy between the Muhlbock mouse and the standard mammal with its spontaneous luteal phase. This warrants some speculation. Is it possible that in the latter there is normally a considerable portion of the pars distalis which functions somewhat independently of the hypothalamus, maintaining a relatively high level of LTH secretion? By this analogy, that portion in close vascular relationship with the median eminence, the zona tuberalis, would then compare with the intact gland of the Muhlbock mouse.

2

Hormones of Pituitary

The pituitary gland was known to Galen but its division into a glandular and a nervous part seems to have been recognized first by the Venetian physician and anatomist G. D. Santorini who wrote in 1724: "Although the infundibulum seems to go straight down, as if into the gland below, and although the anterior part seems to be inserted into it, this is certainly not so. For where it touches the anterior gland, it curves back caudally... ."

Typical neurohypophyseal tissue is found only in vertebrates. The neural complex of urochordates has been considered homologous to the vertebrate hypophysis and neurosecretory material has been found in the cerebral ganglion of ascidians. Olsson and Wingstrand (1954) have suggested that the infundibular organ in *Amphioxus* may be "the functional and perhaps also the morphological counterpart of the hypothalamo-hypophysial neurosecretory system in vertebrates" but little attempt has so far been made to identify vertebrate-like hormones.

In the cyclostome *Petromyzon*, the neurohypophysis consists merely of a thin layer of epithelium and nerve endings in intimate contact with the meta-adenohypophysis. The pituitary gland of Myxinoidea is much reduced, and the hypophyseal and neural elements do not come into close contact. Herlant (1954) has suggested this as the reason why the adenohypophysis in *Myxine* fails to differentiate. Elasmobranchs have a neuro-intermediate lobe, i.e., neural and intermediate lobe tissue interdigitate. Similarly, in teleosts the neurohypophysis usually takes the form of anastomosing strands which penetrate deeply into the adenohypophysis though there are exceptions. The pituitary of the Dipnoi is strongly reminiscent of that of amphibians. Wingstrand (1956), for example, has shown that the neurohypophysis of *Protopterus* is almost identical with that of the urodele *Necturus*. This lung fish is therefore the most primitive vertebrate with a true "neural lobe" but whether this morphological feature has functional significance (an interesting question in view of its estivating habit) remains to be seen.

In amphibians the neurohypophysis is differentiated into two parts in a manner which remains characteristic for all tetrapods. It consists of a median eminence which receives its blood supply from a hypophyseal portal circulation and a neural lobe (or infundibular process) with a separate arterial blood supply and venous drainage. The neural lobe is relatively large in terrestrial amphibians and small in aquatic animals like *Necturus*. The morphology of the pituitary gland varies somewhat in different groups of reptiles, birds, and mammals but the differences are superficial.

All neurohypophyses contain the terminals of axons which originate in cell groups of the anterior hypothalamus and form the hypophyseal tract. In fishes and amphibians there is essentially only a single group of cells, the nucleus preopticus, situated on each side of the third ventricle. In reptiles and in the warm-blooded vertebrates there are two distinct cell groups, the paraventricular nucleus—which remains in much the same position as the preoptic nucleus and is probably phylogenetically the older structure—and the supraoptic nucleus.

Much effort has been expended on the problem of whether the pharmacological actions of mammalian posterior pituitary extracts are due to one or several hormones and whether these active principles occur in the pituitary of other vertebrate classes. Thanks mainly to the pioneer work of Herring (1913, 1915) and Hogben and de Beer (1925) it was soon established that neurohypophyseal extracts of "lower" vertebrates produce the same pharmacological effects as those of mammalian posterior pituitary preparations. This survey has since been almost completed in the sense that neurohypophyseal extracts of one or more representatives of all vertebrate classes have been investigated by a variety of qualitative biological tests. However, it was realized long ago that such results show at best only that related but not necessarily identical principles are present in these extracts. In fact, quantitative assays and calculations of potency ratios indicated that there are differences between the hormones of some vertebrate classes.

Important work during the second decade of the century had shown that mammalian posterior pituitary extracts can be separated into two active fractions. It culminated in the demonstration by du Vigneaud and his associated (1953a,b) that these activities were due to two cyclic octapeptides. Methods similar to his, together with a variety of pharmacological assays are now being applied to the analysis of non-mammalian neurohypophyseal extracts and have recently yielded important results.

Another subject which has much exercised the minds of investigators is that of the site of formation of the "neurohypophyseal" active principles. Earlier workers found it difficult to believe that the neural lobe which cytologically does not have the conventional structure of a gland elaborates endocrine principles. But this position was revolutionized by the discovery of Bargmann and Scharrer (1951) that the neural lobe is only a part of a neurosecretory system, and by their concept that it serves mainly as an organ of hormone storage and release.

Mammalian Neurohypophyseal Hormones

Standardization

The most recent (3rd) International Standard for Posterior Pituitary has been renamed International Standard for Oxytocic, Vasopressor, and Antidiuretic Substances. It consists of acetone-extracted dry powder prepared from fresh bovine posterior pituitary lobes. The unit of the oxytocic, vasopressor, and antidiuretic principles is defined as the activity of 0.5 mg of the dry powder, but it is recognized that the last two activities are properties of the same compound. Assay methods admitted by most national pharmacopoeias are those using the isolated rat uterus and the blood pressure-lowering effect in cockerels for the estimation of the oxytocic hormone (oxytocin), and the pressor or antidiuretic action in anesthetized rats for the estimation of the antidiuretic hormone (vasopressin).

The present standard has many drawbacks both for research and therapy. For instance, the antidiuretic activity of the most commonly used commercial preparation of vasopressin may be due to a peptide which differs from that extracted from ox glands and which may have a different ratio of pressor to antidiuretic potency when injected into human beings. Two preparations standardized by the pressor assay may therefore have very different antidiuretic activities. The adoption of the chemically pure hormones as standard preparations is undoubtedly desirable but seems at present to be difficult because of the relative instability of pure arginine vasopressin.

Chemistry

It was found quite early that the pressor-antidiuretic and oxytocic activities of mammalian posterior pituitary gland extracts can be differentially extracted and precipitated, fractionally adsorbed, or separated by electrophoresis. However, although very active preparations were obtained and the strong suggestion emerged that the active substances were polypeptides, none of these methods led to the isolation of the pure hormones. This was achieved only when Craig's method of countercurrent distribution became available to du Vigneaud and his co-workers.

The oxytocic component (oxytocin) was found to be composed of eight amino acids and ammonia. Further studies of du Vigneaud and his associates and Tuppy (1953) established the amino acid sequence of oxytocin as

CyS · Tyr · Ileu · Glu(NH_2) · Asp(NH_2) · CyS · Pro · Leu · Gly(NH_2)

and du Vigneaud proposed the structural formula which was confirmed by the successful synthesis of the hormone. When subjected to

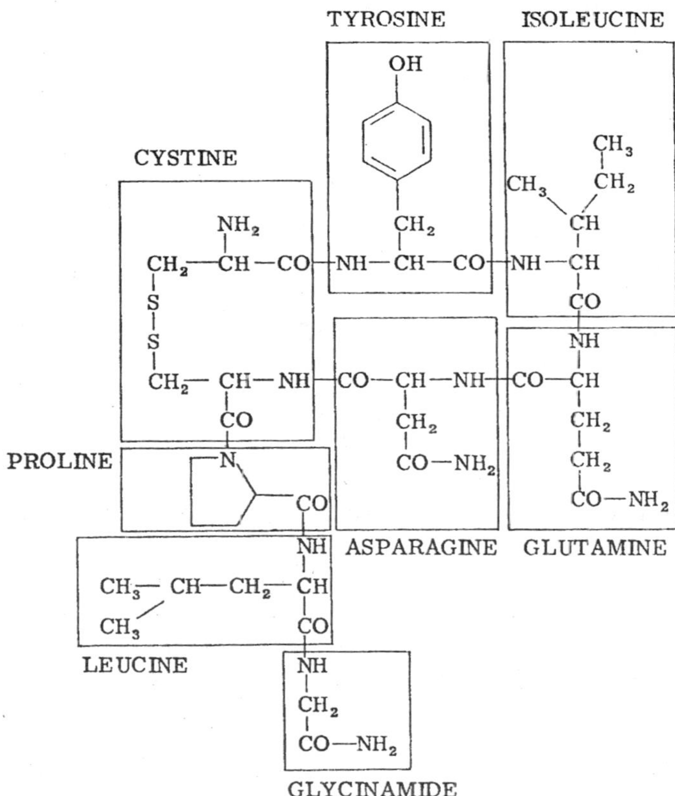

Fig. 2.1. The structural formula of oxytocin.

countercurrent distribution, the biological activity of the synthetic compound was found to be concentrated in a single peak. Its potency was indistinguishable from natural oxytocin when assayed by the chicken depressor method, it had the expected potency when tested on the isolated rat uterus, it was effective on the human uterus *in situ* and produced milk ejection in women when injected intravenously in a dose of 1 μg.

Chemically also, no difference could be detected between the synthetic and the natural polypeptide. Oxytocin synthetized by a different method and compared by further biological tests could likewise be shown to be identical with the natural product. Oxytocin is thus an octapeptide composed of a pentapeptide ring containing cystine, one-half of the cystine moiety possessing a free amino group, and the carboxyl group adjacent to the latter joined to the amino group of tyrosine. The other half of the cystine residue is connected through its amino group to aspartic acid and linked through its carboxyl group to a side chain consisting of the tripeptide Pro·Leu·Gly. It will be seen that all the other neurohypophyseal hormones whose structure is known at present follow the same pattern.

Purified by similar methods as the oxytocic compound, the pressor-antidiuretic fraction of ox pituitary extracts yielded a product with a potency of about 400 U/mg. When analyzed by the technique of Moore and Stein the material yielded eight amino acids, six of which, namely cystine, tyrosine, proline, glutamic acid, aspartic acid, and glycine had also been found in oxytocin. In addition, the hydrolyzate contained phenylalanine and arginine. Further work on beef vasopressin revealed the amino acid sequence

$$\text{CyS} \cdot \text{Tyr} \cdot \text{Phe} \cdot \text{Glu(NH}_2\text{)} \cdot \text{Asp(NH}_2\text{)} \cdot \text{CyS} \cdot \text{Pro} \cdot \text{Arg} \cdot \text{Gly (yH}_2\text{)}$$

a finding which was quickly confirmed by Acher and Chauvet (1954). A structure analogous to that of oxytocin could be postulated and was proved by synthesis.

Similar investigations on the pressor-antidiuretic factor in pig pituitary extracts showed however that this active principle differed chemically from beef vasopressin; it contained lysine instead of arginine in the penultimate position of the peptide chain. A number of mammalian posterior pituitaries, including that of man, have been analyzed since. But although arginine vasopressin was found in some species and lysine vasopressin in others, it seems at present that these are the only vasopressins which mammals elaborate.

Pharmacological Activity of the Pure Mammalian Peptides

The exact potency of pure oxytocin and the vasopressins as measured by the appropriate bioassays cannot be stated at present since even the purest preparations available may contain contaminants. This is suggested by chromatographic results which showed often more than one "spot" when material like du Vigneaud's highly purified arginine vasopressin and commercially available synthetic oxytocin (Syntocinon) were investigated. The source of these contaminations may have been manifold. Du Vigneaud (1955) has commented on the instability of arginine vasopressin, and a solution of the highly purified hormone, particularly when heated before refrigeration, may therefore contain degradation products. Similarly, it cannot be excluded that preparations of a synthetic octapeptide contain traces of the nonapeptide from which it has been obtained by oxidative cyclization. Inactive dimers of the active monomeric molecule may also be present.

It is obvious that, at least theoretically, any of the contaminating peptides may act as antagonists by receptor competition. Synthetic analogs of naturally occurring active peptides which have an inhibitory action in certain bioassays are in fact known. It is, however, likely that more often than not, the small amounts of contaminating peptides are virtually inert. Berde et al. (1961) compared the potency of a crude solution of synthetic oxytocin with a more highly purified preparation obtained by 1095 transfers in countercurrent distribution and found little difference in the ratio of activities. This suggests that, which summarizes the activities of oxytocin, arginine vasopressin, and lysine vasopressin, gives a fair approximation of the potencies of the pure hormones, although the potency ratios are probably more reliable than the absolute values.

It is not particularly surprising that compounds which are chemically as closely related as oxytocin and vasopressin share many qualitative actions. Thus oxytocin has some pressor and antidiuretic action in rats, and the vasopressins have a weak effect on the uterus. The pressor potency of lysine vasopressin assayed in the rat against the International Standard has been estimated as 65-75% of that of arginine vasopressin. It could also be shown that in equipressor doses the antidiuretic effect of lysine vasopressin may differ from that of its arginine analog. Injected intravenously into dogs, lysine vasopressin exerts a much weaker inhibitory action in terms of both intensity and duration of response. In man as well the antidiuretic effect of lysine vasopressin seems to be 3-4 times weaker than that of arginine vasopressin. The difference in the antidiuretic potency of the two vasopressins in rats is more difficult to assess since antidiuresis measured as the maximum depression in urine flow may be much the same. However, the duration of action of lysine vasopressin in this species is usually considerably shorter. The neurohypophysis of dog, man, and rat stores arginine vasopressin. It is therefore of great interest that lysine vasopressin in the pig which synthesizes this peptide is antidiuretically as potent, if not more potent, than the arginine compound.

Structure-Action Relationships of Neurohypophyseal Hormones and Some of Their Synthetic Analogs

The synthesis of oxytocin and of the vasopressins, and the advances in the methods of peptide synthesis generally, led to the preparation of a number of analogs of oxytocin and vasopressin with the hope of gaining some insight into the relationship between the chemical structure and the biological activities of the neurohypophyseal hormones. Apart from yielding some information of this kind these studies had also the surprising result that at least one oxytocin analog (8-arginine oxytocin or vasotocin) was synthesized which was subsequently found to occur in nonmammalian pituitary glands.

With the rapid increase in the number of naturally occurring and synthetic octapeptides there is an obvious need for an agreed nomenclature for this group of substances. Berde and his associates have proposed to number the amino acid residues of oxytocin from 1 to 9. The number of the residue which is replaced is then indicated as a superscript to the "new" amino acid(s). In their terminology arginine vasopressin for example becomes phenylalanyl^{3}-arginyl^{3}-oxytocin, or phe^{3}-arg^{3}-oxytocin. Du Vigneaud and van Dyke (1960) however prefer to reserve the adjectival form of amino acids (e.g. glycyl) for compounds in which an amino acid has been added in peptide linkage. They also suggest that the number which defines the position of the substituent should be placed before the amino acid. Thus oxytocin in which the penultimate residue has been replaced by arginine, would be defined as 8-arginine oxytocin when expressed in the unabbreviated form. Since their procedure is more in line with common chemical usage, it has been adopted in the present article. The problem remains whether all the substances of this type should be expressed as derivatives of oxytocin or whether "laboratory" names like vasopressin, vasotocin, ichthyotocin, etc. should be used in parallel. In view of their historical connotations and failing the decision of an international body, it may be best to retain these terms for the present.

If the two cysteine moieties in the oxytocin or vasopressin molecule are not changed, but the position of the remaining amino acids in oxytocin and the two vasopressins is altered, 1,728,720 different polypeptides are possible. The number becomes even more astronomical when iso-derivatives and amino acids not present in the naturally occurring mammalian hormones are introduced. But although the pharmacological activity of only about 30 synthetic analogs has been tested, certain tentative conclusions about relationships between chemical structure and biological activity can be drawn.

Replacement of one amino acid residue of oxytocin may result in compounds with considerable biological activity (e.g., 2-Phe-, 3-Phe-, 3-Leu-, 3-Val-, 8-Ileu-, 8-Val-, 8-Arg-, and 8-Lys- oxytocin) though none of these peptides have the potency (per molecule) of the naturally occurring mammalian hormone. Other peptides of this type (e.g., 2-Ser-, 3-Tyr-, 3-Try, 5-Glu-, 4-Isoglu-, and 5-Isoasp- oxytocin) have little if any oxytocic action and their other biological activities are also negligible. Note that substitution of isoleucine or valine for leucine in position 8 impairs oxytocic potency only little, and insertion of arginine or lysine decreases oxytocic activity per unit weight of peptide but enhances the pressor-antidiuretic activity in relation to the oxytocic activity, probably because the resulting compounds carry the same tripeptide side chain as do the naturally occurring vasopressins. Introduction of leucine or phenylalanine into position 3 has the effect of decreasing absolute oxytocic potency and increasing (relative) pressor-antidiuretic activity. In contrast, substitution of valine for isoleucine in position 3 leads probably to a "purer" oxytocic substance, 3-valine oxytocin apparently having relatively less pressor-antidiuretic activity than oxytocin. It also seems that positions 4 and 5 in the pentapeptide ring may be critical since the isoglutamine and isoasparagine isomers of oxytocin were found to be devoid of biological action. The pentapeptide ring of oxytocin itself, however, appears to have some slight pharmacological activity.

Inspection shows further that introduction of the same amino acid at different positions may produce very marked quantitative differences in action. A comparison of 2-phenylalanine oxytocin with 3-phenylalanine oxytocin, for example, demonstrates that it is not the presence of the phenylalanine residue in the disulfide ring as such which is mainly responsible for the vasopressin-like effects; the 3-phenylalanine compound is about 90 times as active as the 2-phenylalanine analog. Jaquenoud and Boissonnas (1961) have shown quite recently that substitution of valine in position 8 of oxytocin impairs uterus-stimulating potency per molecule much less than substitution in position 3.

Berde et al. (1961) pointed out that in the group of synthetic oxytocin analogs in which two amino acids residues had been replaced, those which contain the tripeptide side chain of lysine vasopressin have some vasopressin-like activity. Similarly, the triple-substituted oxytocins may be regarded as lysine vasopressin with one amino acid replaced. The weak pressor action which these compounds exert is therefore likely to be due to their resemblance to the naturally occurring hormone.

An indication of the importance of the hydrogen atom on the imino nitrogen of the peptide bond between leucine and glycinamide has been obtained by substitution of sarcosinamide for glycinamide. The resulting compound 9-sarcosine oxytocin showed only a small fraction of the activity of oxytocin itself but the five biological actions tested were lowered to different extents.

Even more interesting is another recently prepared oxytocin analog, desamino-oxytocin—i.e., a peptide which lacks the free amino group on the cystine residue in position 1. It was shown to have a considerably higher uterus-contracting activity (684 ± 32 U/mg) and avian depressor activity (733 ± 23 U/mg) than the purest preparation of oxytocin; its galactobolic potency was also high (400 ± 8 U/mg). The rat pressor activity was lower than that of oxytocin, namely 1.1 ± 0.1 U/mg, but the rat antidiuretic activity appears to be significantly higher (14.9 ± 2.1 U/mg). These results show clearly that the free amino group in oxytocin is not required for most of the biological activities tested.

Several analogs have recently been found which inhibit individual effects of the naturally occurring hormones. Thus 2-*O*-methyltyrosine oxytocin strongly inhibits the uterus-stimulating effect of oxytocin *in vitro*, the mode of action being that of a competitive antagonist. Oxytocin analogs with the peptide chain extended at the amino end by an additional amino acid residue or peptide sequence, such as glycyl-oxytocin or leucyl-glycyl-glycyl-oxytocin also act as inhibitors of oxytocin in several preparations. Moreover, the avian depressor activity of these analogs is distinctly protracted, as is their antidiuretic and oxytocic effect *in vivo*.

Neurosecretion and the Ultrastructure of the Hypothalamo-Neurohypophyseal System

Oxytocin and the vasopressins occur not only in the neural lobe but also in other parts of the hypothalamo-neurohypophyseal complex.

Lederis (1961) has found that the active principles extracted from the hypothalamus of man, sheep, ox, rabbit, rat, and pig behave chromatographically like "glandular" oxytocin and vasopressin, and that the pressor-antidiuretic principle in pig hypothalamus extracts behaves like lysine vasopressin, whereas that in hypothalamic extracts from the other species behaves like the arginine compound. It is, however, by no means clear in what form the peptides are secreted and stored. Nor is it certain whether they circulate in the blood as free or bound compounds, or in what form they are excreted in the urine.

The granular material, demonstrable with selective stains (e.g., Gomori's chrome-alum-hematoxylin-phloxin) in the cell bodies, axons, and terminals of supraoptic and paraventricular neurons, is considered to contain the active peptides. This is suggested (a) by the parallel distribution of the neurosecretory material and biological activity, (b) by the observation that experimental conditions which deplete the neurohypophysis of hormones also empty the neurosecretory vesicles, and (c) by the more recent findings that subcellular particles can be isolated by differential centrifugation of rabbit posterior pituitary homogenates which, with regard to their size and other electron microscopical characteristics, resemble the neurosecretory vesicles seen *in situ*. The isolated subcellular elements were shown to contain 80% or more of the hormonal activities of the undifferentiated homogenate.

The presence of protein in the neurosecretory granules was first stressed by Schiebler (1951). Barrnett and Seligman (1954) and Sloper (1954, 1955) found subsequently that material with the exact distribution of the chrome-alum-hematoxyphil granules could be demonstrated with methods devised to show protein-bound cystine or cysteine. This material could be removed by tryptic digestion. These histochemical results are in good agreement with evidence of another kind which suggests that the hormones in the neurohypophysis are bound to protein. Rosenfeld (1940) concluded from sedimentation experiments in the ultracentrifuge that the vasopressor and oxytocic activities in untreated press-juice of beef posterior lobes were large protein molecules. Van Dyke and his associates (1942) isolated a protein (molecular weight of about 30,000) from beef posterior pituitary glands which had pressor, antidiuretic, and oxytocic activities in the same ratio as an acetic acid extract of the gland. It appeared to be pure as judged by its solubility curve and its Schlieren patterns in the ultracentrifuge. When the oxytocin and the vasopressins had been isolated the question arose whether these peptides represented active fragments of a large molecule connected, for example, by disulfide linkages to longer peptide chains, or whether they were bound to a protein carrier in a less intimate fashion, as e.g., by adsorption or electrostatic forces. Since the oxytocin in crude beef posterior extracts can be obtained by electrodialysis in a 100% yield, and since this treatment does not involve the rupture of peptide or disulfide bonds, the latter possibility seems more likely.

Similarly, Acher and Fromageot (1957) could show that processes which do not involve hydrolysis, such as dialysis against dilute acetic acid, electrodialysis, precipitation with trichloracetic acid, and countercurrent, distribution dissociate the oxytocic and pressor activities from "van Dyke's protein." Acher and his co-workers have recently prepared a protein (neurophysin) from ox, pig, and horse posterior pituitaries which adsorbs oxytocin and vasopressin preferentially and which they therefore regard as the physiological carrier of these hormones.

Whether the active peptides reach the circulation bound to the specific carrier protein cannot be decided at present. Material stainable by Gomori's method has been found in the neurohypophyseal vessels of the giraffe, the rat, and the dog. Heller and Lederis (1962), in electron-micrographs of the rabbit pituitary, have recently noted structures within capillaries which look like the hormone-carrying vesicles or granules, but this was seen so rarely that it may well represent an artifact.

Storage in the Neural Lobe

Neurosecretory material and hormonal activity are present in several parts of the hypothalamo-neurohypophyseal system but in those forms which possess a neural lobe, this structure must be regarded as the main site of storage, both in terms of abundance of neurosecretory material and of hormone content. The amounts of active peptides relative to body weight vary considerably from one mammalian species to another and there is the strong suggestion of a connection between the relative size of the neural lobe and the occurrence of hibernation or desert life. This would agree with the report that the pituitary of the desert-living kangaroo rat (*Dipodomys merriami*) contains more antidiuretic hormone than that of normal laboratory rats although the latter arc 5-6 times heavier.

The neural lobes of all mammalian species investigated may safely be regarded as storage organs since the quantities of active peptides which can be extracted are in all instances many times larger than the minimum active dose or the amounts which normally circulate. In the rat, for example, whose posterior pituitary may contain 1000 mU antidiuretic activity, about 3-15 μU vasopressin/100 gm body weight arc secreted per minute. It can be calculated from this estimate and from the constant obtained by Ginsburg and Heller (1953) in the equation for the plasma clearance of antidiuretic hormone that the corresponding concentration of the hormone in the blood would be 0.8-3.2 μU/ml or 5-19 μU in the total blood volume. The latter figure is of the same order as the minimum (intravenous) antidiuretic dose.

The ratio of vasopressor-antidiuretic to oxytocic activity in extracts of the International Standard powder is, by definition, one. But this ratio is not, strictly speaking, the same as the ratio of the amounts of pure peptides in the gland. First because the vasopressins have some intrinsic oxytocic and oxytocin some pressor-antidiuretic activity, and secondly because it has been shown that the acetone treatment of ox posterior pituitary powder recommended by the pharmacopoeias extracts some of the oxytocin. Both corrections are small and may be neglected in adult males but not in very young or in lactating animals.

Occurrence in the Hypothalamus

In terms of hormone content of the neural lobe, the amounts of hormonal activity found in the hypothalamus are small. The dog appears to be an exception; pressor activity amounting to up to 20% of the activity in the posterior lobe has been recorded but the oxytocic potency of dog anterior hypothalamus extracts is as low as that of other species or probably lower. Indeed it is so low that Vogt (1961) has suggested that it may be accounted for by the intrinsic oxytocic activity of arginine vasopressin. However, Lederis (1962) has recently been able to demonstrate oxytocin chromatographically in extracts of dog hypothalami.

It is likely that in most species the nuclei contain both oxytocin and vasopressin although they may contain them in different proportions. Similar results, viz., a distinct preponderance of oxytocin over vasopressin in the paraventricular nucleus (V/O = 0.2) and the opposite in the supraoptic nucleus (V/O = 2.2) have also been shown in the Algerian camel and in the dog.

These results suggest that estimations of V/O ratios in extracts of the whole hypothalamus are of limited importance since they may derive from the mean of two regions with very different hormone concentrations. Furthermore, they raise several possibilities: (a) that individual neurons produce both hormones but do so in different proportions; (b) that each neuron synthesizes one hormone only but that the proportion of oxytocin-producing neurons and of vasopressin-

Table 2.1. Hormone content and vasopressin: oxytocin ratio (V/O) in the mammalian hypothalamus

Hormone content of hypothalamus			*V/O in*	
Species	*Vasopressin*	*Oxytocin*	*Hypothalamus*	*Neural lobe*
Man	4.6	5.9	1.9	1.6
Macaque	0.4	—	—	—
Dog	20.0	2.3	17.0	1.5
Cat	—	—	1.0	1.3
Ox	0.6	0.5	1.7	1.4
Camel	0.4	1.2	1.1	3.3
Sheep	0.5	0.6	1.9	1.8
Pig	0.7	0.6	2.6	1.6
Rabbit	1.9	1.4	4.0	2.9
Rat	1.4	1.0	2.3	1.6

producing neurons varies in the supraoptic and paraventricular nuclei; (c) that—perhaps in some species only—all the neurons of the supraoptic nucleus synthesize vasopressin and all the neurons of the paraventricular nucleus produce oxytocin, and that the oxytocin found in the former and the vasopressin in the latter are only contaminants present because of incomplete separation of the two nuclear regions at dissection.

Distribution in Mammals

The identification of neurohypophyseal hormones can be attempted by several methods. The only fully convincing procedure consists in the careful purification by chemical and physicochemical means of the unknown active peptide followed by amino acid analysis and establishment of the amino acid sequence. This method has the disadvantage that relatively large amounts of material are needed. Only the glands of a few large and easily obtainable mammals have therefore been analyzed in this manner.

Chromatography and comparison with pure reference substances offer a more limited but still very useful method of identification. It has been used widely and requires only small quantities of active material. With the method of Reindel and Hoppe (1954)—which is more suitable for staining of cyclopeptides than the more commonly used ninhydrin—Heller and Lederis (1958) were able to visualize less than 1 μg of the neurohypophyseal hormones on paper chromatograms. Considerably smaller quantities still (less than 1 mU or about 0.002

μg of oxytocin or arginine vasopressin) could be eluted from areas parallel to reference spots, and biologically assayed.

Chromatographic results, however, may not invariably differentiate between one active octapeptide and another. It was found, for instance, that in the system *n*-butanol-acetic acid-water, oxytocin had the same R_f as one of its synthetic analogs, 3-phenylalanine oxytocin. Similarly, an oxytocic peptide found in teleost pituitaries, which was subsequently shown to differ substantially from oxytocin, appeared in the same position as the latter in both paper and column chromatograms. It follows that the neurohypophyseal active peptides cannot be identified by R_f values only. But even when very small amounts of hormone are available, they can be further characterized, after chromatographic separation, by estimating the potency ratio in bioassays on several organs or tissues.

A combination of chromatography and multiple bioassays has proved very useful in determining the vasopressin elaborated by the hippopotamus; only three glands were available. Vasopressin was separated from oxytocin by paper chromatography and was found in the region of highly purified pig lysine vasopressin. The action of eluates containing the hippopotamus vasopressin were then compared with that of arginine and lysine vasopressin on the isolated hen uterus. It was found that, like lysine vasopressin, the hippopotamus hormone had a much weaker effect. In assays on the water diuresis of unanesthetized dogs the hippopotamus hormone again behaved more like lysine vasopressin than like the arginine analog. Thus both pharmacological methods of comparison and the chromatographic findings showed that the vasopressor-antidiuretic principle in the posterior pituitary of the hippopotamus was not arginine vasopressin and was very probably lysine vasopressin.

In mammals an antidiuretic hormone other than arginine vasopressin has thus far only been found in species belonging to the two surviving groups of the Suiformes which are a suborder of the Artiodactyla or even-toed ungulates. All other eutherian glands and the neural lobes of the three marsupials investigated apparently contained arginine vasopressin. It is of particular interest that the arginine compound has also been tentatively identified in the monotreme *Tachyglossus* since this may be taken to indicate that "arginine vasopressin made its appearance very early in evolution, either in the reptilian stock from which mammals descended or in a very early mammalian ancestor common to monotremes, marsupials, and placental mammals." Since, so far as we know, in phylogenetically recent reptiles 8-arginine oxytocin (vasotocin) is the main antidiuretic hormone and vasopressin appears only in mammals and possibly in birds, it is tempting to connect its occurrence with the ability to produce a hypertonic urine, though the elongation of the loop of Henle was probably an equally or even more significant evolutionary step.

The oxytocic activity in mammalian neural lobe extracts can be accounted for by oxytocin and by the weak intrinsic uterus-stimulating effect of the vasopressins. There is at present no indication that mammalian glands contain an oxytocic hormone other than oxytocin in any considerable quantity.

Significance of Variation in the Chemical Composition of the Vasopressins

Whatever its mechanism, it appears that the process of peptide and protein biosynthesis allows a certain latitude in the incorporation of amino acids. Imprecision of peptide assembly is presumably determined and limited by the resemblance of some amino acids to each other. For example, the three-dimensional structure may be similar (as in the case of valine and isoleucine) or the similarity could include electrical properties (note e.g., the similarity of the pK_1 values for arginine and lysine). The biological activity of new analogs may therefore be expected to vary according to the degree of alteration in their spatial structure, that is to say the new compound will have a better or worse fit to the receptors and will be a more effective or less effective "hormone."

There is much evidence available which suggests that the composition of proteins may be a relatively direct expression of gene structure. We may therefore infer the existence of one or several determining genes for each hormone and, in turn, that changes in the hormone molecule are basically due to gene mutation. Such mutations may be advantageous; it is, for example, conceivable that in the mammalian environment the vasopressins are more effective than the nonmammalian antidiuretic hormone(s). If so, the establishment of vasopressin in the heredity of the mammals would be easy to understand. However, recent studies of structure-action relationship in synthetic analogs of the neurohypophyseal hormones have shown that substitution of a single amino acid may deprive the hormone molecule completely or almost completely of one or more of its characteristic actions. A mutation of this type could obviously lead to extinction of the strain through the development of a lethal gene.

There seems to be a third possibility. A compound may be formed which is of no obvious adaptive value and which is in fact less active than the original molecule, but which permits the carrying organism to "get by"—it is known from other examples that a mutation which confers no benefit or may even be detrimental can become established. When applying this concept to the occurrence of lysine vasopressin in the pig and perhaps other Suiformes, it will be recalled that lysine vasopressin was found to have a weaker antidiuretic action than the arginine compound in the three arginine vasopressin-producing species (man, dog, rat) which have so far been tested. The potency ratio (arginine vasopressin to lysine vasopressin) varies from test species to species but it may be assumed that lysine vasopressin was sufficiently potent to permit the survival of the first mutants of the Suiformes. That lysine vasopressin injected into pigs is antidiuretically as active or more active than arginine vasopressin may be explainable by gradual adaptation of the receptors.

It has already been pointed out that not only the domestic pig but apparently also the hippopotamus synthesizes lysine vasopressin. On the other hand, all the other Artiodactyla so far investigated (ox, camel, sheep, deer) seem to carry the arginine analog. A systematic investigation of the main subdivisions of the Artiodactyla, i.e., of the Suiformes, the Tylopoda, and the Ruminantia is, however, highly desirable in order to put the distribution of the vasopressins, in this order at least, on a firmer experimental basis.

The surviving Suiformes or pig-like animals comprise the Suina or true pigs, the Tayassuinae or peccaries, and the Hippopotamidae. The relationship of these groups is controversial. Many paleontologists have considered the peccaries exclusively New World suids; their evolution in North America is fairly well known from Oligocene times. But although some contest the strictly American origin of this group and assume a common ancestry of suids and tayassuids from the Eocene Dichobunoidea, others regard the peccaries as quite distinct from the Old World Suiformes constituting perhaps a third line in the evolution of artiodactyls, different from the true pigs and hippopotamus, and the ruminants. The origin of the Hippopotamidae is likewise debated. They are assumed either to stem from suids or from late Tertiary Anthracotheres, but the occurrence of lysine vasopressin in *Hippopotamus amphibius* somewhat favours the former assumption. Sawyer (1961) has recently investigated undifferentiated extracts of the pituitary of the collared peccary and has presented strong pharmacological evidence that these glands contained a mixture of arginine vasopressin and oxytocin. An investigation of the chromatographically separated and partially purified hormones obtained from the neurohypophyses of collared peccaries (*Pecari tajacu*) and white-lipped peccaries (*Tayassu pecari*), presently in progress in the writer's laboratory, shows further that both arginine vasopressin and lysine vasopressin may occur in the same neural lobe. Both vasopressins may also occur in a single gland of the African wart hog (*Phacochoerus aethiopicus*) or lysine vasopressin may be present alone.

Thus the investigation of the phyletic distribution of the vasopressins is not only likely to assist taxonomic problems from a rather unusual angle but it carries also the fascinating promise of gaining some insight into the paleontology of a large group of mammals by a biochemical study of recent forms.

Physiological Significance of the Neurohypophyseal Hormones in Mammals

Water and Salt Metabolism

Differences in renal function between mammalian species are small but not unimportant. It is well known for example that the glomerular filtration rate (GFR) of dogs changes rapidly in response to alterations in diet or salt intake whereas that of human beings is remarkably stable. Strauss (1959) has tried to explain this difference on the basis of feeding habits. He points out that the ancestors of the dog "having hunted until they achieved a kill, gorged themselves to satiety thereby presenting the excretory mechanism with a large load of nitrogen and a surplus of sodium (the latter derived from the extracellular fluid of their prey). The ancestors of man, if they resembled present day anthropoids, were largely herbivorous and spent five or six hours daily in feeding, thus hardly ever ingesting more than a minimum of sodium." Unfortunately this attractive interpretation ceases to be acceptable when extended to other species. The GFR of the omnivorous rat responds to protein or salt loading in much the same manner as that of the dog, and the variable glomerular function of the herbivorous rabbit has long been the bane of renal physiologists. It cannot, however, be excluded that species differences in the effects of the posterior pituitary hormones on electrolyte excretion are connected with the degree of lability of renal blood flow and filtration rate. But the main physiological effect of the pituitary on renal function, namely the increase of tubular water reabsorption under the influence of the antidiuretic hormone (vasopressin) appears to be the same in all mammalian species.

Evidence for the physiological role of vasopressin in the regulation of the metabolism of water derives from several sources: (a) Severe interference with or destruction of the hypothalamo-neurohypophyseal complex gives rise to diabetes insipidus. (b) Injection of a small amount of vasopressin or inclusion of the head in a heart-lung-kidney preparation restores the concentrating ability of the diabetic kidney, (c) Dehydration increases the excretion in the urine of an antidiuretic substance which has the chemical and pharmacological characteristics of vasopressin.

Verney (1947) has shown that a small increase of plasma osmotic pressure, produced by the intracarotid injection of a hypertonic salt solution, causes release of vasopressin by stimulating "osmoreceptors" in the vascular bed of the internal carotid artery. Direct proof that the same mechanism operates during the slow process of dehydration is still lacking but should probably only be regarded as a formality.

An exclusively renal site for the regulatory action of vasopressin on the water metabolism of mammals is generally accepted. There is also agreement that (a) water and electrolyte absorption in the proximal renal tubules are not affected by the antidiuretic hormone; (b) the essential effect of vasopressin on the distal segment of the nephron consists in an alteration of the epithelial permeability to water; (c) the production of a hypotonic urine (water diuresis) is mainly due to the decrease of the hormone concentration in the blood arising from the inhibitory effect of the diluted plasma on the osmoreceptors and the very rapid inactivation of the still circulating vasopressin.

However, views on the role of the antidiurctic hormone in the elaboration of a hypertonic urine are to some degree still divided. The "classical" concept may be summarized as: After osmotic equilibration in the thin part of the loop of Henle an iso-osmotic fluid is delivered to a virtually water impermeable segment of the distal tubule from which active absorption of sodium takes place leaving osmotically "free" water. Epithelial permeability to water is then increased under the influence of vasopressin so that water is passively absorbed until isotonicity is again established. The filtrate moves along, most likely to the collecting ducts where the removal of an additional amount of water without solute represents the last phase in the elaboration of a concentrated urine. Smith believed that this final step might be a continuous autonomous process, capable of being carried out in the complete absence of antidiuretic hormone.

The assumption that vasopressin is responsible for reducing the volume of filtrate by facilitating water absorption to isotonicity only, has a special appeal to the comparative endocrinologist since this concept of the renal action of neurohypophyseal hormones could be extended to lower vertebrates (e.g., amphibians) which are unable to prepare a hypertonic urine. Certain experimental results seem, at first glance, to support the independence of the concentrating mechanism from neurohypophyseal function. Shannon (1942) reported that dehydrated dogs with experimental diabetes insipidus and low filtration rates were able to excrete a hypertonic urine, and other workers found that in dogs in full water diuresis, urinary concentration could be raised above the plasma osmotic level when the GFR was markedly lowered. However, the concentrations observed in these experiments were never in excess of 475 milli-osm/liter whereas normal dogs are capable of elaborating a urine as concentrated as 2000 milli- osm/liter. A lowered GFR may thus help in concentrating the urine—it is known that a higher concentration can be achieved by dehydration than by the injection of posterior pituitary extracts into a well hydrated animal—but the antidiuretic hormone remains the most important factor.

More recently the concept that vasopressin acts essentially by increasing the permeability of some tubular segment or segments has been fitted into the countercurrent multiplier theory of Wirz et al. (1951). By this interpretation the hormone increases the diffusion of water from the descending limb of Henle's loop and the collecting ducts. In other words, it permits the specific concentrating mechanism of the mammalian kidney (based on the spatial integration of the renal tubules and blood vessels) to operate in a more efficient manner. However, if the antidiuretic hormone acted simply by allowing diffusion of water, the medullary interstitial space would be progressively diluted and the purpose of the countercurrent mechanism would be ultimately defeated. It has therefore been suggested by Lamdin (1959) and Gottschalk (1960) that vasopressin facilitates not only the transport of water but also that of sodium.

The countercurrent theory of urinary concentration which is receiving increasing experimental support may also aid in the understanding of certain comparative aspects of renal function: (a) the well known differences in the urinary concentration maxima of mammalian species may be based on anatomical divergences in the concentrating apparatus such as the number and length of the thin limbs of Henle's loop; (b) the inability of lower vertebrates to concentrate the urine beyond the iso-osmotic level may not be due to a difference in the mode of action of their antidiuretic hormone but to the morphological characteristics of their kidneys.

As to the "cellular" mechanism of action of the antidiurctic hormone in mammals, it has been suggested in analogy to the effect on the frog skin, that the neurohypophyseal active peptides increase the tubular absorption of water and enhance sodium transport from the lumen by increasing the pore size of the epithelial layer. More recently, Ginetzinsky and his associates have expressed the view that in the tubules vasopressin liberates hyaluronidase which depolymerizes the mucopolysaccharide complex of the intercellular cement so that more water passes through. The last word on this hypothesis has by no means been spoken but some suggestion that hyaluronidase is connected with the production of a concentrated, small volume of urine is provided by the interesting report of Dicker and Eggleton (1960) that after an injection of vasopressin, patients with nephrogenic diabetes insipidus fail to excrete hyaluronidase whereas normal subjects and patients with neurohypophyseal diabetes insipidus excrete large quantities.

High doses of pure oxytocin injected into rats in water diuresis produce a weak antidiuretic effect. Smaller doses may increase urine flow, but this diuretic effect is much more pronounced in rats without an extra water load or after the administration of 0.9% NaCl solution.

Vasopressin has usually a chloruretic and natriuretic effect in dogs and rats, but given in full antidiuretic doses it does not affect the excretion of these ions in man. The increase in the urinary excretion of Cl, Na, and K induced by oxytocin in rats appears to be influenced by the water and salt load of the animals. In dogs oxytocin has little effect on Na or K excretion during water diuresis but greatly increases Na (and sometimes also K) excretion at low rates of urine flow. In man oxytocin has no effect on electrolyte excretion. Since the oxytocin doses which enhance electrolyte excretion have been shown to alter renal blood flow and glomerular filtration in rats and dogs but not in human beings, it may be conjectured that the effect of oxytocin (and probably also of vasopressin) on the urinary output of electrolytes is secondary to vascular changes. Species changes in renal "lability" may perhaps again be invoked. In summary, the conclusion expressed as long ago as 1950 that in mammals the neurohypophyseal hormones play no role of importance in the physiological regulation of the metabolism of Cl, Na, and K can be maintained. It was pointed out at that occasion that the absence of electrolyte disturbances in diabetes insipidus strongly supports this contention.

Cardiovascular System

Administered in large doses, i.e., in quantities much in excess of the minimum antidiuretic dose, vasopressin is known to have a vasoconstrictor effect, including also the coronary and skin vessels.

The physiological significance of these effects is doubtful but, it cannot be excluded that vasopressin, even in very small doses, facilitates the operation of the renal countercurrent system by decreasing renal medullary blood flow. Brooks and Pickford have shown that minimal antidiuretic doses of vasopressin decrease renal plasma flow if it has first been raised by oxytocin. It may also be suspected that in certain pathological conditions, large enough amounts of vasopressin reach the systemic circulation to produce marked vascular effect. Ginsburg and Heller (1953) and Ginsburg and Brown (1956) showed, for example, that the concentration of antidiuretic activity in the external jugular blood of rats anesthetized with ether rose after severe hemorrhage from 10-25 mU to 700-2500 mU/100 ml blood. Similar results have been obtained by Weinstein et al. (1960) in dogs.

Pickford (1961), Lloyd and Pickford (1961a), and Lloyd (1959a, b) reported that in rats the vascular responses to oxytocin and vasopressin vary with the concentration of ovarian hormones in the body. Oxytocin was shown to have dilator effects during diestrus but in the estrous animal, during late pregnancy, or after the administration of ovarian hormone, oxytocin has a pressor and vasoconstrictor action. In normal men or women large doses of oxytocin produce dilatation of skin and striated muscle vessels and a transient fall in blood pressure. However, in contrast to its action in rats, the vasodilation induced by the hormone is maintained in human pregnancy. The pressor action of vasopressin is enhanced in estrous and probably also in pregnant rats.

Whether the changes in the vascular responses of rats to the posterior pituitary hormones are of physiological significance (and are connected with the phasic changes in neurohypophyseal function observed during the estrous cycle in the same species) will have to be further investigated.

Reproductive Organs

Both mammalian neurohypophyseal hormones act on organs concerned with reproduction (uterus, mammary gland), but oxytocin is the more potent one. Present information suggests that oxytocin in mammals may be associated with three aspects of the viviparous habit, viz. internal fertilization, delivery of the young at term, and suckling of the offspring. It seems also to be concerned in the retardation of mammary involution. Firm evidence for the physiological importance of the hormone is available only for the last one of these functions. Beller et al. (1958) for example saw milk ejection in women after an intravenous dose of as little as 10 mU. The amounts of endogenous oxytocin released by suckling seem to be considerably higher. Cross (1961), for example, found in the rabbit that suckling evokes a milk ejection response which resembles the effect of an intravenous injection of 50 mU oxytocin. Inactivation of the posterior pituitary by an electrolytic lesion of the supraoptico-neurohypophyseal tract in the rabbit abolishes the milk ejection reflex; in this species therefore, and in the rat there can be little doubt that natural suckling stimulates the neurohypophysis. An increase in the oxytocic activity of the blood of sheep at the time of milk ejection has been recently reported by Fitzpatrick (1961).

Although an injection of oxytocin produced milk ejection in all mammals so far investigated, it has been pointed out that its physiological significance is likely to vary from species to species. Oxytocin is primarily concerned with the removal of alveolar milk which in the goat, for example, with its capacious udder cisterns, is only a small fraction of the yield at a normal milking. This may account for the reports that goats give almost normal yields under conditions in which the neurohypophyseal milk ejection reflex is inoperative.

Parturition

A whole symposium has recently been devoted to an inquiry into the role of oxytocin in parturition. The problem is befogged by reports that viable young can be delivered despite elimination of the neurohypophysis. In one of the latest papers on this subject Gale and McCann (1961) placed electrolytic lesions in the median eminence of rats at various stages of gestation, thus producing severe diabetes insipidus. In rats receiving lesions on days 7-9 and maintaining gestation to term, 32% experienced difficulty during delivery. Lesions placed after day 13 did not impair parturition. The authors conclude that in the rat oxytocin is not essential for parturition. Numerous other observers, however, have recorded prolonged labour and maternal and fetal death under similar conditions in other mammalian species including women. Caldeyro-Barcia and Poseiro (1959) have summarized some of the facts in favour of the assumption that oxytocin participates in the process of human labour. They point out that (a) all the characteristics of the contractions produced by the pregnant human uterus during an infusion of oxytocin are indistinguishable from those in spontaneous labour, (b) the doses of oxytocin needed to produce these effects are of the same order as the minimum antidiuretic dose of vasopressin; (c) a specific enzyme which inactivates oxytocin (oxytocinase) appears in the blood of primates in pregnancy. To be added to this is the demonstration in pregnant animals that stretching of the cervix produces a release of oxytocin, and that, in some mammals at least, the oxytocic activity of the blood increases substantially *intra partum*.

Whether, as some investigators hold, the essence of the role of the neurohypophysis in labour consists in an increase of the uterine reactivity to oxytocin at term or whether an increased release of the hormone is necessary, cannot be decided at present. It may be that species differences in the estrogen-progesterone equilibrium are involved. Likewise, it has yet to be decided whether oxytocin is material in the initiation of labour or whether it merely helps to expel the fetus.

Facilitation of sperm transport by oxytocin

This hypothesis which has been extensively discussed in a number of recent articles derives support from the demonstration in several species that at mating oxytocin is released and uterine motility increases. Massage of the seminal vesicles and ampullae of the ram also releases an oxytocic substance. The biological value of accelerating the transport of the spermatozoa has been questioned, and there seems little reason at present to suppose that fertility would be lower if all the spermatozoa were obliged to spend an hour or more in the uterus.

3

ADENOHYPOPHYSIS IN FETUS

This consideration of the functioning of the developing hypophysis cerebri will be based almost exclusively on experimental studies in the rat which have been made at the University of Minnesota. Accordingly, my presentation concentrates upon the functional relation between the hypophysis and the adrenals, thyroid and gonads. It also includes some results from an unfinished study of growth hormone.

The present observations should be prefaced by a short account of the techniques which have proven of value in enlarging the understanding of endocrinology in the fetus. First, Raynaud (1943) has published a technique for destroying the hypophysis or thyroid of the fetal mouse *in utero* by means of a narrow beam of X-rays. Second, Wells (1946a, 1950a) devised a method for "extra-uterinizing" the fetal rat and then subjecting it to surgical adenoprivia and/or repeated subcutaneous injections. Third, Wells (1947) described a procedure for depriving the fetal rat of the hypophysis by decapitating it *in utero*. Fourth, Sethre and Wells (1951) published a method of hypophysectomy of the fetal rat, so that the developing thyroid and its blood vessels are spared.

Almost simultaneously a number of workers published methods for depriving the fetus of the gonads and hypophysis.

HYPOPHYSIS-ADRENAL SYSTEM

There are at least seven lines of experimental evidence indicating that the fetal rat hypophysis produces a corticotrophic hormone (ACTH), that this hormone causes the adrenal cortex to produce an active principle(s), and that the functional relation between the hypophysis and adrenal cortex is reciprocal.

1. Hypophyseoprivia retarded the development and growth of the adrenal. In a fetus lacking the hypophysis for 102 hours, about 19 per cent of the total period of pregnancy, the "adrenal volume/body weight" ratio was smaller (–39%) than that in a littermate control of the same sex. The outer zone of the cortex was abnormally thick. This suggests that hypophyseoprivia slows up the transformation of cells in the outer zone into cells of the deeper zones.
2. Such retardation in the development of the adrenal cortex after hypophyseoprivia was prevented by giving the "hypophysectomized" fetus a series of four or five subcutaneous injections of ACTH. In addition, the injections advanced cellular differentiation in the cortex beyond that found in untreated littermates.
3. Unilateral adrenalectomy in the fetus led to a compensatory hypertrophy of the intact adrenal. The hypertrophy was demonstrated by data on the "adrenal volume/body weight" ratios in the experimental fetuses and in nonadrenalectomized controls. In the enlarged gland the intermediate zone of the cortex (future zona fasciculata) was wider than normal, and its cells were larger than normal.
4. Compensatory hypertrophy of the adrenal was prevented by implanting a pellet of cortisone under the skin of the unilaterally adrenalectomized fetus. The reliability of this observation was supported by a companion experiment in which the compensatory hypertrophy was not prevented by implanted desoxycorticosterone acetate.
5. Subcutaneously implanted hydrocortisone retarded the growth of the fetal adrenal, as did also implanted cortisone. Each of these treatments reduced the volume of the gland, decreased the width of the cortex, thickened the outer cortical zone (future zona glomerulosa), reduced the size of the cells in the intermediate zone and widened the "lipid-free region" between these two zones.
6. Implanted hydrocortisone retarded the growth of the anterior lobe of the hypophysis. The characteristics of this retardation were: a reduced size of the anterior lobe, a poorly-defined pattern of cords of epithelial cells, a reduced diameter of these cells, a scant amount of cytoplasm in the cells, a distortion of some of the nuclei, a scarcity of mitoses, and a reduced size and number of sinusoids.
7. Bilateral adrenalectomy of the fetus seemed to induce an enlargement of the anterior hypophysis and an enhanced development

of the hypophyseal cells and sinusoids. This observation was based, however, on only three cases.

These several observations are taken to mean that the hypophysis-adrenal system in the fetus is independent, or largely independent of any hormones from the maternal hypophysis or from the placenta. This view is supported by two observations: (a) that maternal hypophysectomy did not modify the growth of the fetal adrenals, and (b) that it did not influence appreciably the development of the fetal hypophysis. The view is not necessarily discredited by the hypothesis that in the hypophysectomized pregnant monkey a placental corticotrophin may be a factor which prevents the maternal adrenals from undergoing as much retrogression as that found in the hypophysectomized nonpregnant monkey.

Hypophysis-Thyroid System

We have six lines of experimental evidence that in the rat the hypophysis-thyroid system begins to function before birth.

1. Hypophyseoprivia retarded the histological development of the thyroid follicle. This change was observed consistently when the period of hypophyseoprivia was as long as three days, but was not large enough to be demonstrated volumetrically by the paperweight method.
2. The retarded development of the follicle after hypophyseoprivia was prevented by a series of 4 subcutaneous injections of TSH. Incidentally, the injections stimulated a degree of development of the thyroid which exceeded that observed in untreated littermates.
3. Injected thyroxin or triiodothyronine (TIT) slowed up the development and growth of the thyroid. The changes included a reduced volume of the thyroid and a slowed development of the follicles in the thyroid.
4. The changes usually induced by thyroxin or TIT were prevented by simultaneously-injected TSH. The "thyroid volume/body weight" ratios were considerably larger than those in uninjected littermates.
5. Thyroidectomy in the fetus seemed to speed up the cellular development of the anterior hypophysis. The manifestations of this acceleration, presently recorded only in a dissertation, centered around the number of basophilic epithelial cells, the amount of cytoplasm in these cells and the granularity of the cytoplasm.
6. Injected thyroxin seemed to produce the opposite effect, a retarded cellular development of the anterior hypophysis.

All these observations are interpreted to mean that the hypophysis-thyroid system in the fetus is independent of hormones from the maternal hypophysis or from the placenta. This view is supported by three observations: (a) that thyroidectomy in the pregnant female did not modify the growth of the fetal thyroid; (b) that hypophysectomy in the gravid female rat did not induce any significant morphological change in the fetal thyroid; and (c) that the injection of radioactive iodine into fetal rabbits was followed by a smaller concentration of radioactive iodine in the thyroid of an "hypophysectomized" fetus than in the thyroid of a control fetus.

Hypophyseal Gonadotrophin

Experiments were designed to test the hypothesis that the developing hypophysis cerebri produces a gonadotrophic hormone. The results, especially those in male fetuses, support this hypothesis.

Seventeen fetuses were deprived of the hypophysis by removing the head *in utero* at the stage of one to four days before expected delivery. Five of the seventeen were given a series of five subcutaneous injections of equine gonadotrophin. All fetuses and littermate controls were killed near term, at the age of 21 days and 15 hours.

In males subjected to hypophyseoprivia, the main effect was a reduction in the number and size of the interstitial cells of the testis. Such reduction was prevented by injections of gonadotrophin.

Growth Hormone

A study of growth hormone is not yet finished. The results up to now seem to indicate that the fetal hypophysis contains a growth-promoting principle which normally stimulates the growth of the fetus.

Hypophyseoprivia was produced surgically *in utero* in fetuses of 18 days and 20 hours; after a postoperative period of 67 hours, autopsy was performed (21 days and 15 hours). In an individual litter with one or two experimental fetuses, the nontreated littermates of the same sex were used as controls.

Hypophyseoprivia alone retarded the ponderal growth of the fetus. This effect was prevented by subcutaneous injections of bovine growth hormone, but was not prevented by injections of bovine serum albumin.

4

X-Rays and Hypophysis

The general problem of the role played by the fetal hypophysis in the development of the mammalian fetus and of the control of its endocrine physiology has already been the object of several publications and conferences. This report is chiefly confined to the question of the relation between the hypophysis and sexual differentiation in the mouse fetus; and within the outline of this symposium, the comparative examination of the effect of suppressing the hypophysis in this species and in other rodents.

Destruction by X-rays of the Hypophysis of the Mouse Fetus

First attempts to remove the hypophyseal primordium of the mouse fetus by cautery or by decapitation were unsuccessful because of the minute size of the fetus. A special technique of localized irradiation was evolved, which permits one to direct over a minute area of the fetus, after laparotomy of the mother under ether anesthesia, a thin beam of X-ray, through the uterine wall and the fetal membranes. The beam is canalized by metallic tubes (canalizer and localizer) and the localization is done by contact of the extremity of the localizer with the uterine wall, just above the head of the fetus, which is seen by rendering the uterine horn transparent with the help of oblique illumination. In order to avoid displacements due to the respiratory movements of the mother, the uterine horn is immobilized on a special apparatus. The fetus is inclined slightly, head downward so that the beam passes through the head without any possibility of going through the body. The conditions of irradiations were: 33.5 KV; 36 mA; anticathode of Mo; $\lambda = 0.95$ Å; filter, A1 = 0.05 mm; distance to

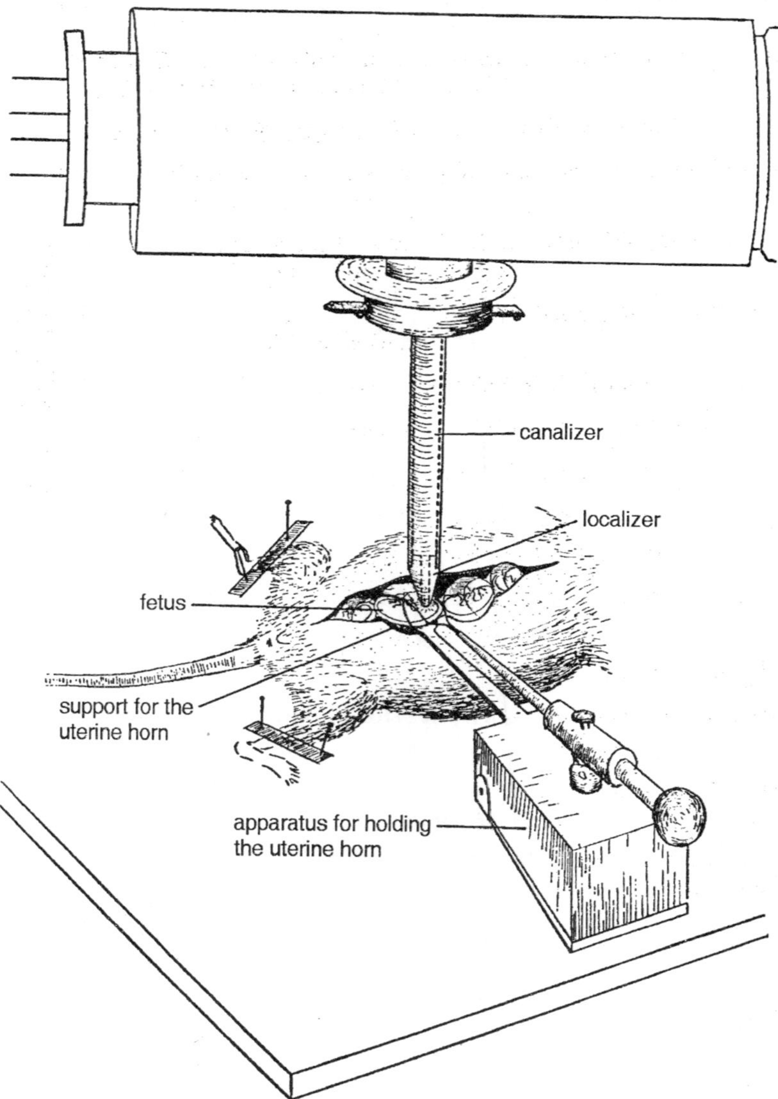

Fig. 4.1. Scheme of the apparatus used for localized irradiation of the mouse fetus.

uterine horn = 6.5 cm; dose rate at the level of uterine horn = 36,000 r/minute (under these conditions a dose of 120,000 r requires only 3´20´´ of exposure). After irradiation the uterine horn is allowed to slip back into the abdominal cavity, and the abdominal incision of the mother is closed. The mother is sacrificed only some hours before

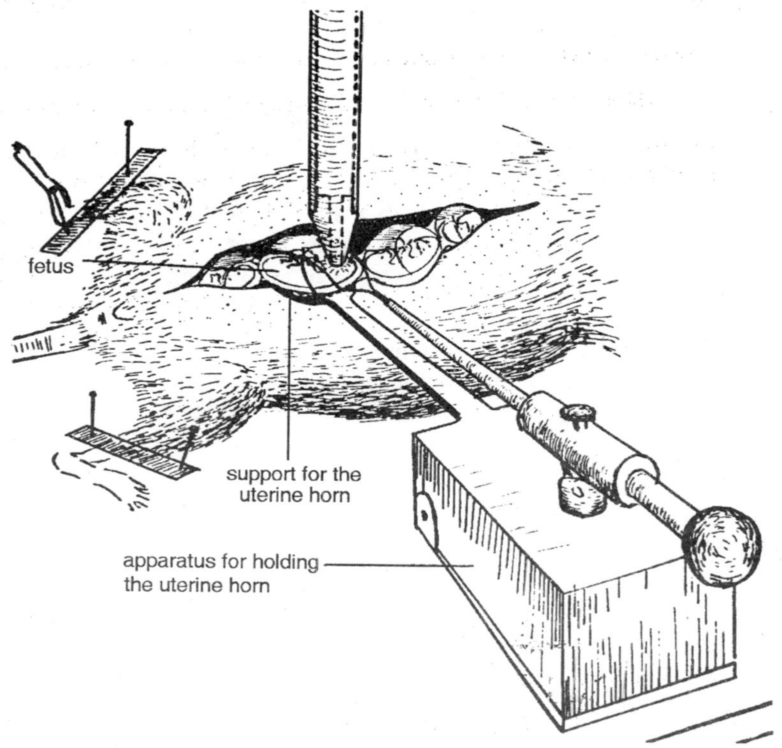

Fig. 4.2. An enlargement of a part of fig. 4.1.

term, and the fetuses are removed, sacrificed, autopsied, fixed, and histologically studied in serial sections.

Three hundred fetuses were irradiated under these conditions at the age of 12 days and some hours. Some among them were again exposed to radiation at 13 days and some hours; the total dose of X-rays, applied at the base of the skull (with a localizer having a circular aperture 2-3 mm in diameter), ranged from 5,000 to 200,000 roentgens. Eighty of these fetuses continued to develop and reached the full term of intra-uterine life.

At 12 days the hypophysis of the mouse fetus displays the structure; Rathke's pouch is closed but still thin and connected with the buccal epithelium; the anlage of pars nervosa is a minute neural evagination.

The irradiated fetuses which have reached term present a great reduction of the head and a necrosis of most of the irradiated tissues. Microscopic study shows that according to dosage and other factors, either partial or total destruction of the hypophysis, associated with lesions of neighbouring tissues exposed to the rays, is achieved. In regard to the condition of the hypophysis, two groups of fetuses are to be considered:

1. Fetuses which have been subjected to only one irradiation at twelve days with a dose of 5,000 to 100,000 r. Under these conditions the neural lobe of the hypophysis is totally destroyed with doses of 40,000 r and more, but Rathke's pouch persists until term as a little mass of cells. With the higher doses in this series, this residue is smallest, and is composed of altered cells looking totally inactive.
2. Fetuses which have been subjected to two consecutive irradiations of the head, the first at the age of 12 days and the second at 13 days. The dose administered at each irradiation was 100,000 roentgens. Under these conditions total destruction of both anlagen (pars nervosa and Rathke's pouch) of the hypophysis has been obtained in two fetuses. In the area where the hypophysis would have been, one can see only fragments of necrotic cells. Almost total destruction is seen in four other embryos (only some altered cells persist at the periphery of a central mass of necrotic cells). Partial destruction is seen in still four other fetuses. In these we found a small residue of Rathke's pouch, formed by overstained cells, whose volume does not exceed 1/1000 of the normal hypophysis at term. It is chiefly on material of this second group that this study is based.

Effects of Hypophysectomy on Sexual Development of the Fetus

At the time of irradiation (12 days) the morphologic sexual differentiation of the fetus is just beginning; in the two sexes gonads still have a similar structure (testes are distinguishable from ovaries in the second half of the twelfth day), and the genital tracts of male and female fetuses are alike; mullerian ducts begin their development and the Wolffian ducts are continuous as far as the cloaca; at 14 days the Wolffian and mullerian ducts are present in both sexes (the latter ducts already display some regression in the cranial part) and the urogenital sinus and external genitalia are not sexually differentiated. It is during the second half of the fifteenth day and during the following stages that sexual characteristics develop in these sections of the urogenital tract. It has been established that early castration of the mouse fetus (destruction of gonads by X-rays) suppresses masculinization of the genital tract and leads to a feminine pattern. In the castrated female fetus, the urogenital sinus and external genitalia develop according to the normal female type. This indicates that secretions of the fetal testis control the masculinization of the fetus; in the female embryo, the development of sexual characters does not seem to be under hormonal control. These facts seem to be in agreement with the monohormonic conception of sexual differentiation of Wiesner (1934, 1935), but the possible intervention of female hormones of extraembryonic origin in the castrated fetuses still cannot be ruled out.

The problem now is to know whether or not the functional activity of the fetal gonads is controlled by secretions of the fetal hypophysis. What is the effect of hypophysectomy on the genital tract of the mouse fetus?

To date, twenty fetuses having received 200,000 r on the anlage of the hypophysis have been studied, some whose hypophysis has been destroyed totally (1♂ and 1♀), almost totally (2♂ and 2♀), or reduced to a small residue of more or less altered cells (8♂ and 6♀). In all these fetuses the sexual differentiation of gonads and of genital tracts proceeds in a normal or almost normal manner until term of intra-uterine life.

Genital Tract of the Male Mouse Fetus Deprived of Hypophysis

After hypophysectomy the testes occupy a normal male position in the abdominal cavity, on either side of the bladder; the gubernaculum (*conus inguinalis*) is well developed. Microscopic examination shows that the testes although slightly reduced in size (a reduction of 1/10 in the length of their axes) display an almost normal structure. Tubules are also slightly reduced in size (their diameter is approximately 9/10 that of the controls) and the number of germinal cells in the tubules is often greatly reduced. The interstitial gland, however, is well developed. In fact, except for one fetus in which it seems somewhat reduced, the interstitial gland of all fetuses whose hypophysis is totally or subtotally destroyed is very well-developed and display a normal aspect. The cells are large and the large eosinophilic cytoplasmic zone characteristic of the normal cell is found near the nucleus. Wolffian ducts are well developed as are also the seminal vesicles. Mullerian ducts have retrogressed. Coagulating glands and lateral and ventral prostatic glands have developed and have reached approximately normal size. The external genitalia are of the male type; the balano-preputial fold describes an almost complete circle. Some deficiencies of masculinization have, however, been noted: in almost all fetuses, the anogenital distance is a little shorter than the control one; in two fetuses the preputial fold is incompletely closed (but variations also have been observed in controls); in one fetus with a little hypophyseal residue a small remnant of the caudal part of the mullerian duct persists, and ventral prostatic buds are absent; a similar although even smaller vestige of mullerian ducts, however, also can be found in the male control (of the same litter) whose ventral prostatic buds are just beginning to develop. The significance of these differences, therefore, cannot actually be evaluated and can be interpreted on the basis of general retarded development in these fetuses as well as on the basis of a deficiency in male hormone stimulation.

Genital Tract of Female Mouse Fetus Deprived of Hypophysis

This tract is similar to the normal female one. The sexual differentiation has proceeded in an almost normal manner. The ovaries are slightly reduced in size (a reduction of 1/10 in the length of their axes, in comparison to control ones) but display an almost normal structure. The ovarian cortex is a little thinner than that of the control, and the number of germinal cells is greatly reduced, but as in controls the germinal cells have entered meiotic prophase and their nuclei show pachytene and diplotene states. The gonoducts have developed normally; mullerian ducts are continuous and have fused in their posterior parts in the genital cord; Wolffian ducts are retrogressed; the urogenital sinus has formed a vaginal cord in its dorsal part; and the anlagen of the external genitalia are of female type; the anogenital distance, however, is a little shorter than normal.

Thus experimental evidence shows that destruction of the fetal hypophysis, in the mouse, has but little, if any, effect on the sexual development of the fetus.

Growth of the Hypophysectomized Mouse Fetus and Condition of some Organs after Hypophysectomy

Growth of the Fetus

Upon first irradiation, the weight of the fetus is only 50 to 100 mg. It reaches 600 to 900 mg at term of gestation (the greater part of the head is destroyed). Measurements taken outside the irradiated field show that growth is only slightly impaired by the destruction of the hypophysis; all body dimensions are reduced by about 10 per cent when compared with the controls.

Thyroid Gland

From the structure of the gland in the hypophysectomized fetus it is clear that after destruction of the hypophysis, the anlage of the thyroid has been able to develop and to undergo morphologic differentiation leading to the formation of vesicles; but the gland is reduced in comparison to that of the control; the vesicles are minute and contain either no colloid substance or only a minute amount of it on the periphery of their lumen. In all probability these deficiencies in the condition of the thyroid gland may be attributed to the suppression of the hypophysis. Owing, however, to the technical procedure used (in the 12-day embryo, the irradiated field lies in the proximity of the pharynx), an uncertainty exists, because of the possibility of deleterious action by the rays upon the cells of the anlage of the gland, or upon the neighbouring tissues in which the epithelial part of the gland is growing.

Adrenal Gland

The adrenals of hypophysectomized fetuses are reduced in size. The surface occupied by the cortex in a median section of the gland, in reference to the surface of medulla in the same section, is reduced in comparison to controls by 20 to 30 per cent in almost all hypophysectomized fetuses. The cortical cells of hypophysectomized fetuses are reduced in size and lack the eosinophilic character observed in controls. The general aspect of the gland is that of hypodevelopment and hypofunction.

Liver

In the liver, sections stained by Best's carmine method show the presence of glycogen but in a lesser concentration than in controls.

Mammary Glands

It is a well known fact that striking sex differences exist in histogenesis of mammary anlage of the mouse. In the female fetus the mammary cord remains connected to the epidermis. In the male the mammary anlage becomes separated from the epidermis. After early radio-destruction of the testes the mammary anlagen of the male fetus develop like those of the female. After injection of male hormone into the female fetus, the mammary buds became detached from the epidermis and isolated in the underlying mesenchyme. Testes and androgenic hormones are responsible, therefore, for separation of the mammary buds from the epidermis. Consequently, the condition of mouse fetus mammary glands can be used as a criterion of male hormone action. In all our hypophysectomized fetuses, the mammary anlagen follow a normal type of histogenesis: in male fetuses mammary anlagen become separated from the epidermis while they remain connected to the epidermis in the female fetuses.

Interpretation of the Effects of Hypophysectomy in the Mouse Fetus

From the results of our investigation, it is clear that after the destruction of the hypophysis by X-rays, sexual development in the mouse fetus proceeds in a normal or almost normal way, in both sexes. In the testes the interstitial tissue is well developed and in the genital tract of the male fetuses, masculinization follows a normal or subnormal course. The state of this tract (gubernaculum, Wolffian ducts, adnexal glands, all well developed) is a proof of normal or subnormal androgenic activity in the testes following destruction of the hypophysis.

Now, the following question arises in connection with the problem of anterior pituitary-gonad relationships in fetal life: do the above findings allow the conclusion that in the mouse fetus, gonadal activity is independent of hypophyseal stimulation? The author feels that such a view is probably untenable for the reasons stated as follows.

1. We are unaware of the exact moment at which total destruction of the fetal hypophysis or arrest of its physiologic activity is produced by the X-rays. Our material was much too limited, due to the difficulty of the operation, to permit a study of results of hypophyseal destruction at different stages in the course of fetal development, Certain facts, however, indicate that destruction probably occurs very shortly after treatment: in some cases no remnants of the two hypophyseal anlagen were found at term. This would indicate their early destruction. Moreover, the dosage used (200,000 r) is sufficiently great to inhibit rapidly all physiologic activity of the irradiated cells (calculations show that in one nucleus, 8×10^6 ion pairs must be produced by one irradiation with 100,000 r). On the other hand, as can be judged from the state of the genital tract of the fetus, a dose of 100,000 r applied to the gonads is sufficient to suppress rapidly the activity of the fetal testis. In all probability we can, therefore, conceive that suppression of fetal hypophyseal function takes place in the first hours following the second irradiation.
2. Administration of a high dose of estrogenic hormone to the pregnant mouse is followed by atrophy of the interstitial gland of the fetal testis, and by an arrest of masculinization of the genital tract.

These results support the idea that activity of the fetal testis is controlled by a gonadotropic secretion which can be suppressed by injection of estrogenic hormone into the pregnant mother.

3. We must think of the possibility of intervention by another source of gonadotropic secretion in the hypophysectomized mouse fetuses: after destruction of the fetal hypophysis, it is possible that placental or maternal gonad-stimulating substances reach the fetus and stimulate the fetal gonads. A suggestion in accord with this line of thought may be found in the observations of Astwood and Greep (1938), who showed that a water-soluble extract of rat placenta is capable of stimulating the ovaries of hypophysectomized rabbits.

It would thus appear, that following destruction of the fetal hypophysis, a gonadotropic substance, perhaps of placental origin, acts on the fetus, stimulating the gonads and causing the fetal testes to produce a masculinizing secretion. Among all the possible interpretations of the above experimental results, this one seems to be the most probable. If such a conclusion is accepted, the formation of a male genital tract in our hypophysectomized male mouse fetuses would be induced by gonadotropic stimulation of extra-fetal origin. Variations in the concentration of this gonadotropic substance would be responsible for the degree of activation of fetal interstitial tissue (which in some instances looks more active than that of the control testes) and for the slight variations observed in the degree of masculinization of the genital tract.

On the other hand, reduction in the number of germinal cells in gonads of irradiated fetuses suggests that irradiation of the hypophysis suppresses a special F.S.H. type of gonadotropic function in the fetal hypophysis. This is a function that placental secretions cannot perform. It is also possible that radiation exerts an indirect effect on the germ cells, perhaps through alterations of the blood, or as a consequence of necrosis of the cerebral cells. Much more experimental evidence will be required before a definite conclusion can be reached on this point.

Concerning the other functions of the fetal hypophysis of the mouse, my research has not been sufficiently extended on this point to lead to definite conclusions. However, the measurements taken on fetuses irradiated on the trunk and the abdominal region indicate that suppression of the fetal hypophysis by X-rays retards but slightly the somatic growth of the fetuses. The atrophic condition of the thyroid and adrenal glands in hypophysectomized fetuses suggests a pituitary-thyroid and pituitary-adrenal relationship in the fetal life of the mouse.

Effects of Hypophysectomy on the Genital tract and Endocrine Glands in Fetuses of other Species of Rodents

Extensive research on the problem of functions of the hypophysis in fetal life have been undertaken on rat fetuses, rabbit fetuses, and hamster fetuses. All this work has been carried out by the decapitation technique, which leads to an immediate suppression of the hypophyseal function. We still do not know, however, whether the decapitation of the fetus does not involve effects other than those of hypophysectomy. For example, ligature of the neck leads to obstruction of the trachea and of the oesophagus. At the same time the stomach must remain empty, and circulation is disturbed.

The effects of hypophysectomy by decapitation of rat fetuses are discussed here by L. J. Wells. Therefore, this work will not be summarized here, but only brief mention will be made of the fact that in rat fetuses Wells (1947, 1950) and a co-worker (1951) have established that the absence of the hypophysis does not cause any marked retrogression of testicular interstitial cells. On the same type of material, Jost and Colonge (1949), and Jost (1951b) noted that the interstitial gland seemed slightly reduced but well developed in the decapitated fetus (decapitation of the fetus at 16 days, microscopic examination at 20-21 days); the operation did not significantly modify the sexual differentiation of the fetus. On the other hand, when the rat fetal testis is grafted onto the seminal vesicle of an adult hypophysectomized host, its interstitial gland is atrophic. It thus appears that the fetal testis is sensitive to the stimulating action of hypophyseal hormones. In the rat as well as in the mouse, it is possible that after fetal hypophysectomy a gonadotrophic secretion of placental origin may reach the fetus and stimulate its gonads.

Rabbit Fetus

The researches of Jost and his collaborators on decapitated fetuses contribute evidence for the existence of physiological relationships between anterior-pituitary, and other endocrine glands (gonads, thyroid, adrenal) in fetal life. On the other hand, the suppression of the hypophysis has at the most a slight effect on fetal growth.

Pituitary-gonad relations

Decapitation of the male fetus at 19 days is followed by a certain number of genital alterations. The testes migrate down into the abdominal cavity, but histologic examination shows that the interstitial gland is distinctly reduced as compared to controls of the same age.

Epididymis and Wolffian ducts are essentially normal but alterations appear in the seminal vesicles, prostatic buds, and external genitalia. Prostatic buds, particularly, show deficiencies with absence or arrest of development in the lateral and anterior buds. The tubal and uterine parts of the mullerian ducts retrogress, but remnants of the caudal parts of the ducts sometimes persist and are more developed than in the control fetuses. Furthermore, external genitalia display alterations: the preputial fold is incompletely closed and the phallus displays hypospadias or a feminine morphology.

From all these modifications it may be inferred that partial arrest of masculinization follows suppression of the fetal hypophysis. This alteration of sexual development implies that the secretion of androgenic hormone by the testis is quantitatively deficient in the headless fetus. This deficiency is attributable to the absence of fetal hypophyseal secretion, since after administration of gonadotropic hormone to the headless fetus, masculinization of the genital tract proceeds normally. Moreover, of four decapitated rabbit fetuses with grafted hypophyses, two showed less important alterations of the genital tract. In these two fetuses, prostatic buds were less reduced in size and in one of them, the external genitalia were of the male type.

Study of the development of these genital abnormalities after decapitation shows their onstart to be after the twenty-second day of intra-uterine life. Stimulation of the testis by fetal hypophysis is thus especially important during days 21 to 23. This correlates strikingly with the fact that in the fetal hypophysis the number of cells showing a MacManus-positive reaction increases greatly between days 19 and 23. Thus there appears to be a correlation between the state of activity of the fetal hypophysis and the chief period of testicular activity in rabbit fetuses. From these considerations it is apparent that there is in the rabbit fetus a critical period (19-24 days) for the development of pituitary-gonad relationships, of the genital tract, and of testicular activity. Similarly, in the rat fetal hypophysis, Jost and Tavernier (1956) have shown that MacManus-positive cells appear in the anterior lobe at two distinct stages: first at 15 days in the anterior and ventral regions, and next at 17 days, in the central zone.

Pituitary-thyroid relationships

In rabbit fetuses decapitated at 18-19 days (either by total decapitation, or by the cutting off, through the mouth, of the upper part of the head only), the thyroid gland pursues its structural development. This gland, which at 18-19 days was composed almost entirely of solid cords of cells, displays an evident follicular organization in the decapitated fetuses sacrificed on day 28. The hypophysis is, therefore, not necessary for the first stages in the development and structural organization of the gland. Soon, however, striking differences appear between the thyroids of decapitated fetuses and those of the controls. In headless fetuses, the amount of follicular colloid is reduced, and many incompletely filled follicles can be observed in the glands However, the existence of numerous vacuoles of resorption, found in the follicles of thyroids from some headless fetuses, is rather surprising, and although this is suggestive of a stimulated gland, it still remains unexplained. Study of fetal pituitary—thyroid relationships by physiological techniques shows that thyroid glands of decapitated fetuses collect less radioactive iodine (I^{131}) than the glands of controls. The differences are particularly marked for the period 22-28 days. Experimental evidence thus suggests that the pituitary is not necessary to the initial organization of this gland, but, that from the twenty-second to the twenty-fourth day's stage on, it controls the intensity of thyroid function.

Furthermore, Courrier and Zizine (1957) have demonstrated that after injection of labelled iodide into the pregnant rabbit, the ratio of erythrocytes radioactivity to plasma radioactivity is greater in the fetuses than in the mother. This ratio decreases after thyroidectomy of the mother. It is possible that this is indicative of a low secretory output by the fetal thyroid.

On the other hand, Jost and Picon (1958) have undertaken a study of the function of the thyroid hormone in the fetal organism. After thyroidectomy of the fetus *in utero* at age 23 days, the lipid content of the body is increased (fetuses sacrificed at 28 days). In thyroidectomized fetuses injected with thyroxine, the lipid concentration decreases and returns to a normal value.

Geloso (1956) has established that the thyroid of the rat fetus is able to synthesize monoiodothyrosine, diiodotyrosine, thyroxine and 3,5,3′ triiodothyronine. The rate of formation of thyroxine is greater in the fetal organism than in the mother. Furthermore, after the injection of I^{131} into pregnant rats thyroidectomized at 19-20 days of gestation, the plasma of their fetuses contains labelled thyroxine. This hormone cannot be detected in the fetuses deprived of their hypophysis; thus the thyroid gland of the fetal rat is able to synthesize and to release thyroxin and this active secretion would be under hypophyseal control. Furthermore, Jost (1957) observed that after injection of an antithyroid substance into the pregnant rat, hypertrophy of the thyroid gland of the fetuses is produced only if the fetal hypophysis is present. This hypertrophy is thus, in all probability, related to hyperactivity of the fetal hypophysis and is suggestive of pituitary-thyroid interrelationships.

In the hamster it is known that the fetal thyroid gland accumulates radioiodine on the thirteenth day of development, the initial appearance of follicles being on the twelfth day. After decapitation of the fetuses, Foote and Foote (1949) have observed a retarded development of the thyroid. The gland acquires follicles, but these remain narrow and show no colloid in their lumina.

Pituitary-adrenal relationships

Jost (1948, 1951a, b, 1953, 1954) studied this aspect in rabbit fetuses. After decapitation of the fetus, the adrenal cortex displays retarded growth with a decrease in its width and in the lipid content of its cells. On the other hand in decapitated rat fetuses, Cohen (1958) observed that the adrenal glands are deficient in ascorbic acid, whereas injection of ACTH under certain conditions restores the concentration of this substance in the gland. It thus appears that the fetal hypophysis controls the different physiological activities of the adrenal gland. This agrees with the extensive studies of Wells and his collaborators on hypophyseal control of the adrenal gland in the rat, before birth.

Glycogen in the liver

This is also, in rabbit fetuses, under the control of hypophyseal and adrenal activities.

5

METABOLIC HORMONES

Metabolism is the total of enzyme catalyzed chemical reactions that are involved in maintaining the structure and functions of an organism. Complex molecules can be constructed from simple building blocks: simple sugars, amino acids, fatty acids and nucleotides. Conversely, complex molecules (proteins, triglycerides, polysaccharides) can be degraded to provide building blocks for synthesis of other complex molecules. The ultimate energy source for all activities of any organism is that stored in chemical bonds. Simple sugars can be degraded to carbon dioxide and water, releasing energy that can be used to drive other processes.

It should not be surprising to learn that essential interconversions of carbohydrates, lipids and proteins as well as the consumption of simple sugars for energy are regulated by hormones. Abnormal metabolism is responsible for the overt symptoms of many endocrine disorders, and by far the most common disorder, diabetes mellitus, is caused by a deficiency in a metabolic hormone, insulin.

Maintenance of any organism involves utilization of energy derived from hydrolysis of chemical bonds. Carbohydrates represent the most immediate energy source for synthesis of adenosine triphosphate (ATP) or other high-energy reserves (such as guanosine triphosphate (GTP) or creatine phosphate). In mammals as in other vertebrates, glucose is the primary carbohydrate energy source. All of the cell's machinery for energy production (oxidative metabolism) relies on hydrolysis of chemical bonds in glucose and glucose metabolites. Glucose and its intermediates may be utilized as precursors in the synthesis of many other essential molecules, including lipids of all sorts, proteins and nucleic acids. It is possible to convert other molecules (amino acids, glycerol, lactate) to glucose, which can be used in the synthesis of ATP or other compounds as needed. Production of glucose from amino acids, glycerol or lactate is termed *gluconeogenesis*.

The major tissues involved in oxidative metabolism in mammals are liver, muscle, fat, nervous and kidney cortex. In addition, red blood cells, the kidney medulla and the testes have low oxidative capacities and produce lactate through anaerobic respiration. Lactate is released into the blood and must be metabolized by other tissues (primarily liver and kidney cortex). Each tissue performs highly specialized functions, and the generalized metabolic schemes for metabolism are modified for the special requirements of each tissue.

Many differences exist in the oxidative and synthetic processes that occur in different tissues, and it is important to understand these differences before discussing endocrine regulatory mechanisms that operate in the intact organism. The homeostasis of energy metabolism can only be appreciated fully when one is aware of the roles played by these tissues as they are influenced by diet, starvation and muscular work. The principal ways in which muscle, liver, adipose tissue and brain metabolism relate to carbohydrate, lipid and protein stores will be examined, as will their regulation under different physiological conditions.

Dietary differences determine to a considerable extent the endogenous mechanisms for metabolic homeostasis. Carnivores eat other organisms that consist primarily of protein and absorb principally amino acids from the gut as well as lesser amounts of fatty acids, glycerol and monosaccharides. Herbivores such as the ruminants have a bacterial flora in their guts that digests cellulose to glucose. The bacteria metabolize glucose and produce a variety of compounds for their own use as well as by-products such as short-chain fatty acids. Fatty acids and amino acids constitute the major oxidative substrates absorbed by the ruminant. Omnivores, such as ourselves, obtain a greater proportion of carbohydrates in their diets than do either carnivores or ruminants. The proportions of carbohydrates, proteins and lipids vary from meal to meal in human diets. The composition of metabolites in hepatic portal blood of animals with these varied diets places different demands on the regulatory mechanisms that must be employed in utilization, conversion, excretion and storage processes. In this chapter general aspects of hormonal influence on carbohydrate, lipid and protein metabolism under differing physiological conditions are considered.

Intermediary Metabolism

The complete pathway for carbohydrate-based energy metabolism is the oxidation of glucose to CO_2 and water, yielding a net increase in ATP. Complete oxidation of glucose involves *glycolysis*, the *tricarboxylic acid* (TCA) or *citric acid* cycle and the *electron transport*

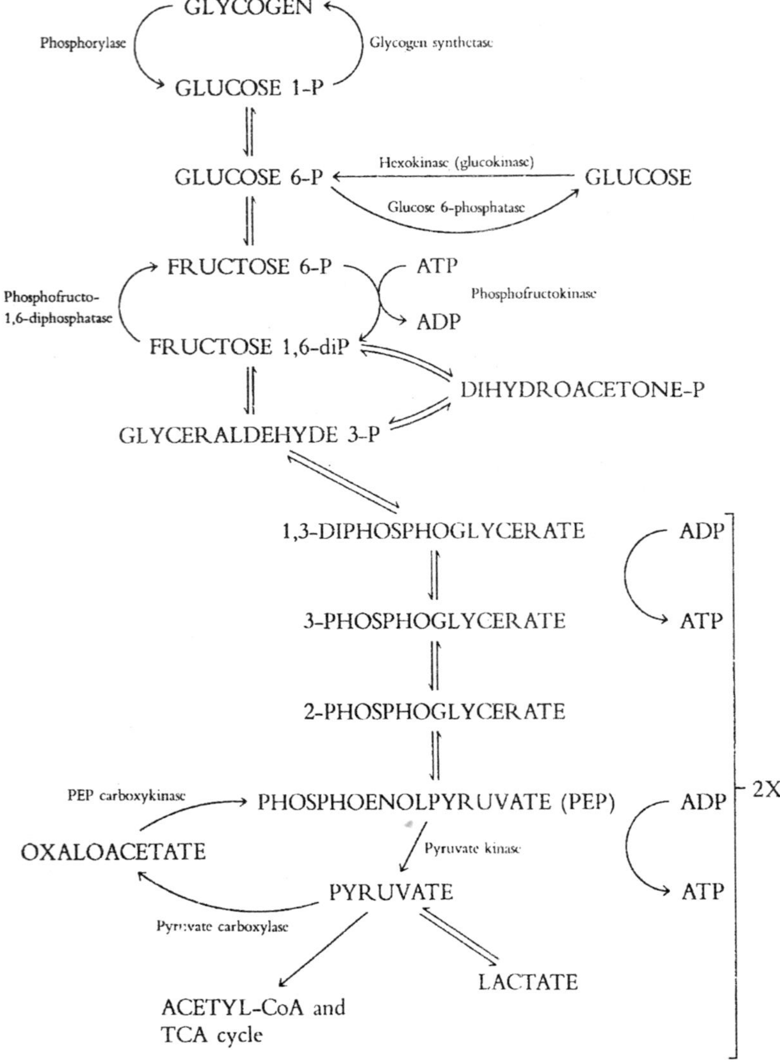

Fig. 5.1. Glycolysis. This scheme summarizes the steps in glucose oxidation through the Embden-Myerhof pathway. P, phosphate.

chain. Generally glycolysis involves stepwise conversion in the cytosol of the six-carbon glucose molecule to two three-carbon intermediates known as pyruvate. A small quantity of ATP is produced by oxidation of glucose to pyruvate. An alternative pathway by which glucose may be oxidized is the *pentose phosphate pathway*, which is closely linked to lipid synthesis. This pathway metabolizes about 30% of the glucose oxidized in the liver cell and provides reduced compounds (NADPH) necessary for fatty acid synthesis, for inactivation of steroids and for drug detoxification. Some of the pentose sugar produced may be utilized

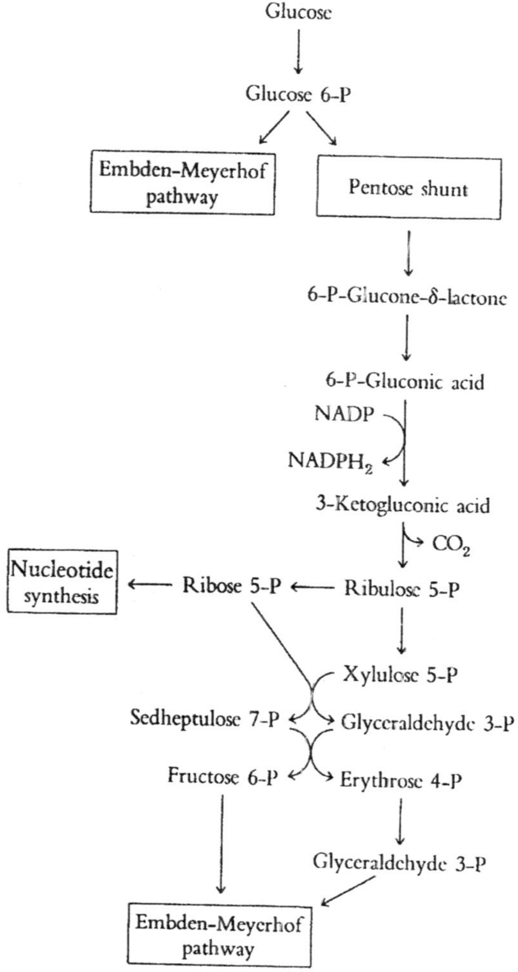

Fig. 5.2. Major steps in the pentose shunt.

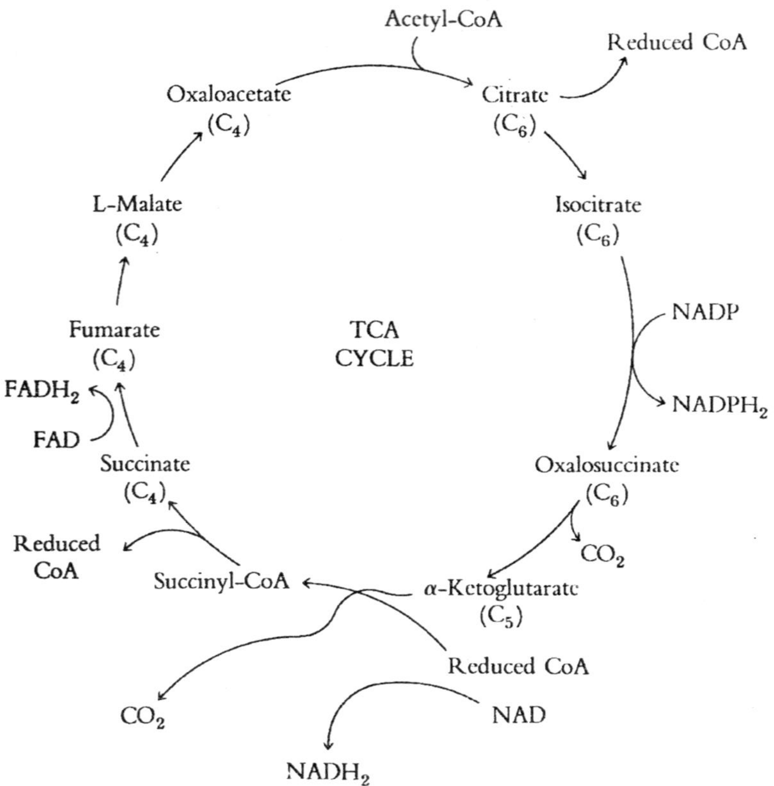

Fig. 5.3. The tricarboxylic acid (TCA) cycle.

for ribose and deoxyribose synthesis and in formation of nucleotides for later incorporation into nucleic acids. In the TCA cycle, pyruvate enters the mitochondrion where it is oxidized to CO_2 and acetate in the form of acetyl-coenzyme A (acetyl-CoA). Acetyl-CoA is further oxidized to yield more CO_2 and a quantity of special reduced compounds (such as NADH). Electrons are transferred from these reduced compounds sequentially through a chain of electron acceptor-donor molecules known as the electron transport chain. At each transfer, an electron drops to a lower energy state while releasing a small quantity of energy, some of which can be trapped in the synthesis of ATP. The final electron acceptor is oxygen (O_2), which, when reduced by acceptance of electrons, forms water. If oxygen is not available (i.e., under anaerobic conditions), this sequence is altered by conversion of pyruvate to lactate. Neither the TCA cycle nor the electron transport chain of the mitochondrion functions under anaerobic conditions.

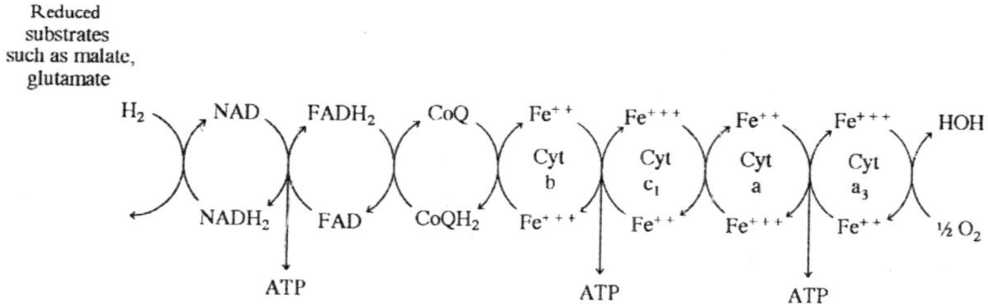

Reduced substrates such as malate, glutamate

H_2 — NAD / $NADH_2$ — $FADH_2$ / FAD — CoQ / $CoQH_2$ — Fe^{++} Cyt b Fe^{+++} — Fe^{+++} Cyt c_1 Fe^{++} — Fe^{++} Cyt a Fe^{+++} — Fe^{+++} Cyt a_3 Fe^{++} — HOH / ½ O_2

ATP ATP ATP

Fig. 5.4. The cytochrome system.

Glycolysis

There are three limiting steps in glycolysis, each of which is catalyzed by a unidirectional enzyme (that is, the reaction may be catalyzed only in one direction because of thermodynamic considerations). Free glucose enters a cell by facilitated diffusion but cannot be utilized by that cell until it has been phosphorylated. This step is catalyzed by a unidirectional enzyme, *hexokinase* (most tissues) or *glucokinase* (liver). Phosphorylated glucose may be polymerized and stored as glycogen (*glycogenesis*), which constitutes a temporary reserve of glucose within the cell, or it may undergo glycolysis (including the pentose phosphate pathway). Hydrolysis of glycogen (*glycogenolysis*) is accomplished by the enzyme complex known as the *phosphorylase* system.

The second limiting step in glucose metabolism involves the unidirectional enzyme *phosphorylase fructokinase* (PFK), which is

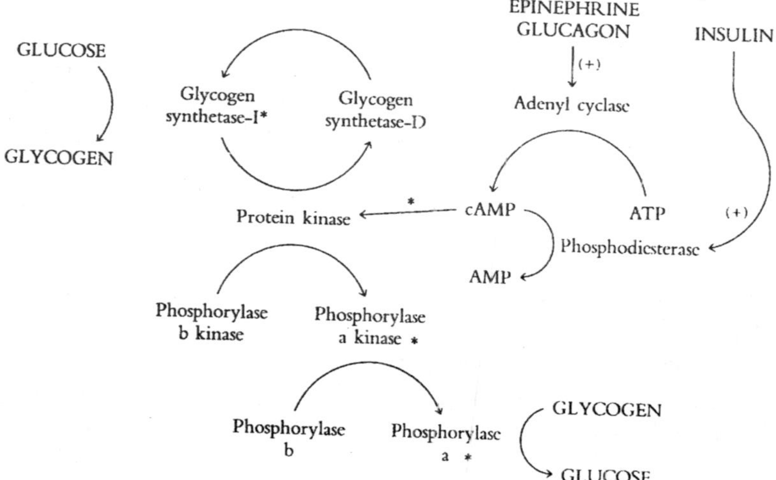

Fig. 5.5. Glycogenolysis and glycogenesis.

responsible for converting fructose 6-phosphate (derived directly from glucose 6-phosphate) to fructose 1,6-diphosphate. This step is essentially irreversible and commits the cell to oxidize glucose to pyruvate. The presence of substrate (fructose 6-phosphate) activates PFK, and this reaction is considered to be the major rate-limiting step in glycolysis.

The final limiting step in glycolysis is the irreversible conversion of phosphoenol pyruvate (PEP) to pyruvate by the enzyme *pyruvate kinase*. Sufficient energy is released upon hydrolysis of PEP to synthesize ATP directly.

Tricarboxylic Acid Cycle and Electron Transport

The TCA cycle is regulated primarily through the availability of acetate for combining with coenzyme A to yield acetyl-CoA. Either glycolysis through pyruvate or oxidation of fatty acids provides acetyl-CoA for TCA-cycle metabolism. The availability of certain TCA-cycle intermediates such as oxaloacetate may also influence operation of the TCA cycle. Such intermediates may be synthesized from amino acids. Utilization of the reduced compounds (for example, NADH) generated by the TCA cycle also determine its continued operation. For example, the unavailability of oxygen as the final electron acceptor will not allow utilization of reduced compounds produced by the TCA cycle, causing the system to halt with an accumulation of pyruvate. Increase in this substrate determines the conversion of pyruvate to lactate by lactate dehydrogenase (an enzyme that catalyzes this reaction in either direction) depending upon relative availability of substrates.

Lipogenesis and Lipolysis

Synthesis of fatty acids is accomplished essentially by combining units of acetyl-CoA to form fatty acyl-CoA molecules. This synthesis requires NADPH produced in part through efforts of the pentose phosphate pathway. Esterification of fatty acids (linkage to glycerol via ester bonds) involves coupling of three long-chain fatty-acyl-CoA units one at a time to glycerol phosphate to form monoglycerides, diglycerides and finally triglycerides (neutral fats). Triglycerides may be stored or, as in the case of the liver, be coated with lipoprotein and secreted as droplets into the blood. Cholesterol synthesis also involves acetyl-CoA as a precursor and is accomplished in the liver.

Lipids are transported in the blood in several ways. Small fatty acids are released directly into the blood where they are referred to as free fatty acids or *nonesterified fatty acids* (NEFA). The latter name is more suitable than "free" because NEFAs have low solubility in aqueous medium and must form a complex with serum albumin for

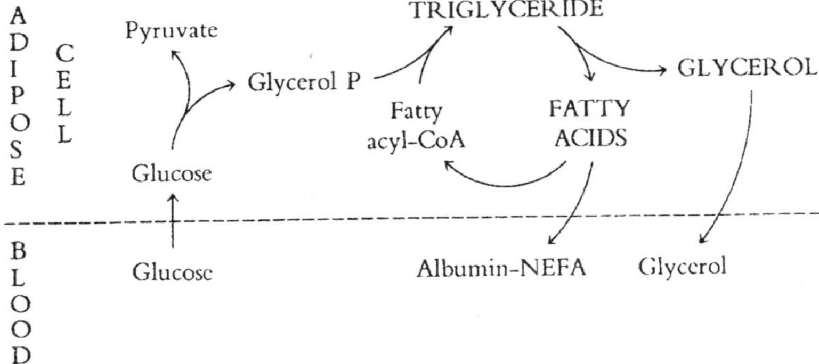

Fig. 5.6. The triglyceride-fatty acid cycle in adipose tissue.

transport. Larger fatty acids are transported as triglycerides (esterified: attached by ester bonds to glycerol). The liver packages triglycerides and releases them in small protein-coated droplets called *very low density lipoproteins* (VLDL). A small amount of cholesterol is also present in VLDLs. Hydrolysis of VLDLs including the triglycerides occurs in capillary beds of adipose and certain other tissues by the activity of *lipoprotein lipase*. Fatty acids and glycerol are released and can diffuse into the adjacent cells.

Larger transport droplets called *low density lipoproteins* (LDL) contain about 60% of the cholesterol that occurs in human blood. These LDLs are about 2 nm in diameter and consist of about 1500 cholesterol esters surrounded by a phospholipid-cholesterol layer and coated with about 20 molecules of apoprotein B and some minor apoproteins. The apoprotein acts as a binding protein for receptors located in pits on the surface of various cellular types. The nature of the apoproteins determines to what cellular type a lipoprotein droplet can bind. The LDL is internalized to complete the transfer of cholesterol from the liver. Steroidogenic tissues utilize LDLs as an important source of cholesterol for synthesizing androgens, estrogens, progesterone, corticosteroids and vitamin D_3.

High density lipoprotein (HDL) droplets transport cholesterol from tissues to the liver where it is metabolized and excreted. About 30% of the blood cholesterol in humans is found in HDLs, and HDLs may provide some cholesterol for steroidogenesis. In rats, HDLs transport most of the cholesterol. There appears to be a relationship between HDLs (especially one type known as HDL_2) and the risk of heart attack. Normal HDL levels are 45 and 55 mg/dl in blood of men and women, respectively. Higher risk of heart attack is correlated with lower levels of HDLs. A great deal is yet to be learned about HDL levels, metabolism and health before we can draw firm conclusions from these observations. For example, what is the meaning of the high HDL levels reported for runners or higher levels (80-100 mg/dl) observed in alcoholics?

Fatty acid mobilization involves hydrolysis of triglycerides by the enzyme *triglyceride lipase* releasing a fatty acyl-CoA and diglyceride. The diglyceride is hydrolyzed by diglyceride lipase to yield another fatty acid plus monoglyceride. Monoglyceride lipase hydrolyzes the latter to yield glycerol and fatty acids. Fatty acids may be metabolized through the β-oxidation process to acetyl-CoA, which can be used to synthesize ATP, or they may be secreted directly into the blood through which they travel to another tissue for oxidation or use in resynthesis of fat. Glycerol similarly may be converted to glycerol phosphate in the same tissue or released into the blood. Glycerol phosphate may contribute to resynthesis of glucose (gluconeogenesis) or may again participate in fat synthesis.

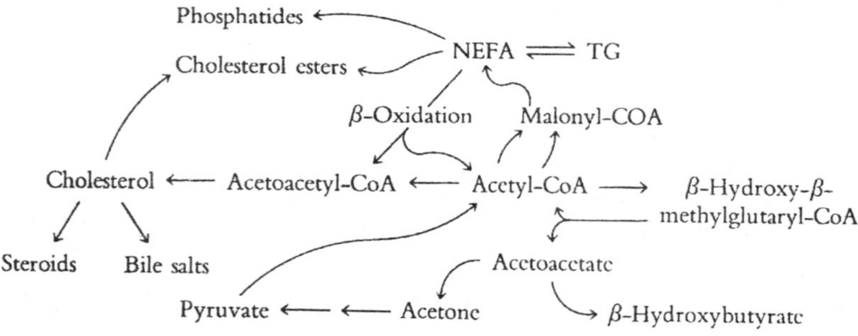

Fig. 5.7. β-Oxidation of fatty acids.

Gluconeogenesis

Gluconeogenesis occurs primarily from amino acids, but glycerol or lactate may also provide gluconeogenic substrates. Dietary or tissue proteins must first be hydrolyzed to amino acids, which may in turn be deaminated via transaminase reactions to yield glucose intermediates (ketoacids) plus ammonia. The ammonia is generally combined with CO_2 via the urea cycle to form urea for excretion. This process is extremely important in carnivores, which assimilate large quantities of amino acids from their diets and relatively little carbohydrate.

The process of gluconeogenesis is essentially a reversal of glycolysis but must utilize different pathways because of the unidirectional nature of hexokinase (glucokinase), PFK and pyruvate kinase. Alanine, for

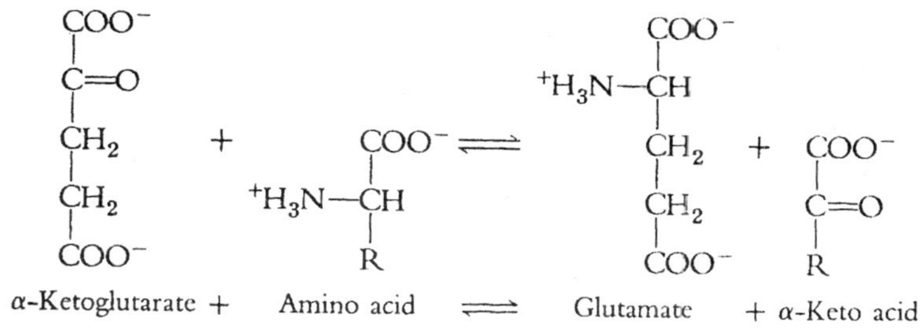

$$\begin{array}{c}COO^-\\|\\C{=}O\\|\\CH_2\\|\\CH_2\\|\\COO^-\end{array} + \begin{array}{c}COO^-\\|\\{}^+H_3N{-}CH\\|\\R\end{array} \rightleftharpoons \begin{array}{c}COO^-\\|\\{}^+H_3N{-}CH\\|\\CH_2\\|\\CH_2\\|\\COO^-\end{array} + \begin{array}{c}COO^-\\|\\C{=}O\\|\\R\end{array}$$

α-Ketoglutarate + Amino acid $\rightleftharpoons$ Glutamate + α-Keto acid

Fig. 5.8. General transamination reaction.

example, may be converted directly to pyruvate through transamination. Pyruvate may then be converted to oxaloacetate by a unidirectional enzyme, *pyruvate carboxylase* and ATP. Other amino acids (except for leucine) may be transaminated to various glucose intermediates and find their way through pyruvate or oxaloacetate, or both, to glucose. It should be noted that transamination reactions require an ammonia acceptor; in this case, as for alanine, the acceptor is α-ketoglutarate. In liver mitochondria the resulting glutamate can be readily deaminated back to α-ketoglutarate by *glutamic acid dehydrogenase* with the ammonia being transferred directly into the urea cycle.

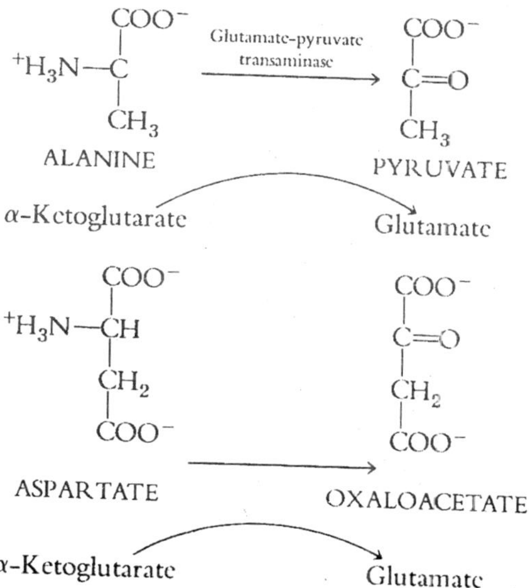

Fig. 5.9. Transamination of alanine to pyruvate and of aspartate to oxaloacetate.

Oxaloacetate is converted to PEP by another unidirectional enzyme, *PEP carboxykinase*, and GTP. Increased levels of PEP influence resynthesis of fructose 1,6-diphosphate, which is directed to glucose 6-

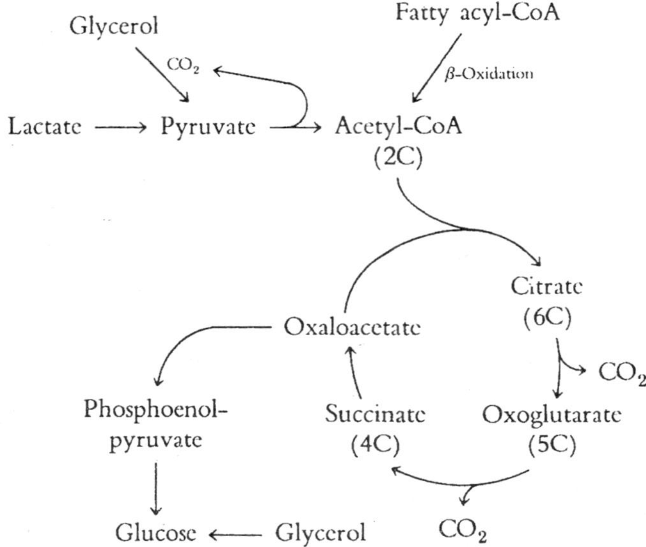

Fig. 5.10. Metabolism of pyruvate or acetyl-CoA via the TCA cycle.

phosphate by a third unidirectional step catalyzed by *fructose 1,6-diphosphatase*. Finally, the last unidirectional enzyme in the gluconeogenic scheme, *glucose 6-phosphatase*, produces free glucose that can diffuse from the cell into the blood. However, this final step does not occur in most tissues and is unique to those tissues that participate in regulating circulating glucose levels, that is, liver and kidney cortex.

Production of glucose from amino acids, like the complete oxidation of glucose, involves both cytosol and mitochondrial events. Pyruvate produced in the cytosol via glycolysis may enter the mitochondrion or may be converted by pyruvate carboxylase to oxaloacetate while still in the cytosol. Mitochondrial membranes are not permeable to oxaloacetate, but mitochondrial oxaloacetate produced in the TCA cycle or through transamination of amino acids may be converted first to other intermediates, such as malate, which may diffuse into the cytosol and may be reconverted to oxaloacetate. This mitochondrial shuttle was proposed to account for observations in the rat and some other species in which PEP carboxykinase was found only in the cytosol, yet transamination to oxaloacetate occurred within the mitochondrion. In rabbits, however, PEP carboxykinase is present in mitochondria, whereas in the guinea pig the enzyme is found in both compartments. These observations emphasize that caution must be used in generalizing about cytological localization of specific metabolic events in gluconeogenesis when considering different species.

Lipids may contribute to gluconeogenesis following hydrolysis of triglycerides to fatty acids and glycerol. Glycerol may be phosphorylated to glycerol phosphate and converted to a three-carbon intermediate in glycolysis. Glycerol phosphate can be oxidized to acetyl-CoA or can be a source of glucose via the gluconeogenic pathway. Since fatty acids inhibit glycolysis, it is more likely that glycerol metabolism will follow the gluconeogenic pathway. Fatty acids through oxidation to acetyl-CoA do not constitute any net synthesis of glucose intermediates. Although oxaloacetate can be synthesized in the TCA cycle, the two carbons of acetyl-CoA will be lost as CO_2 and cannot end up in the glucose molecule. An accumulation of acetyl-CoA in liver results in formation of ketone bodies, which enter the blood and are metabolized by other tissues or excreted.

Lactate under aerobic conditions is converted to pyruvate by lactate dehydrogenase and can be directed toward glucose synthesis through the gluconeogenic pathway. However, no new net synthesis of glucose occurs in this manner since glucose probably was the original source of the lactate. Muscles may release lactate following vigorous contraction when sufficient oxygen is not available. Several other tissues, for example, red blood cells, normally respire anaerobically and contribute lactate for gluconeogenesis. The liver and kidney cortex metabolize lactate that has entered the blood from other tissues.

Metabolic Events of Particular Tissues

Liver

The mammalian liver is capable completely of (1) removing from or releasing glucose to the blood, (2) storing of glucose as glycogen, (3) metabolizing glucose to CO_2 and water with attendant production of ATP, (4) synthesizing triglycerides from fatty acids and glycerol for storage or transport as lipoprotein droplets to other tissues, (5) hydrolyzing triglycerides to fatty acids that may be released or further oxidized to acetyl-CoA for ATP production, (6) synthesizing other lipids or ketone bodies and (7) performing gluconeogenesis from amino acids, glycerol or lactate precursors. In addition, there are myriads of other synthetic activities performed by the liver, including synthesis of a wide range of proteins (renin substrate, somatomedins, serum albumin, etc.), phospholipids and urea, as well as conjugation of steroids and detoxification of many drugs, including alcohol.

In omnivores and in carnivores the major dietary carbohydrate is glucose. Ruminants take in little carbohydrate through their diets. The most important metabolic role for the liver of omnivores, ruminants and carnivores is to provide adequate circulating oxidative fuels during periods of acute or prolonged fasting and to process and direct excessive fuels to adipose tissue for storage as fat. The role differs in relation to the sources of energy in the diet. The liver is anatomically situated ideally for this role since about 70% of its blood supply comes directly through the portal vein, which drains the gut and contains the absorbed products of digestion as well as insulin from the endocrine pancreas. In humans the liver rapidly removes excess glucose by a facilitated diffusion mechanism that is augmented by the enzyme glucokinase, which has a high affinity for glucose and converts it to glucose 6-phosphate. Unlike the hexokinase that phosphorylates glucose in other tissues, hepatic glucokinase is not inhibited by accumulation of product (glucose 6-phosphate), and uptake of glucose is maximized. Cell membranes are impermeable to phosphorylated forms of glucose, and only four metabolic avenues are open to it. Glucose 6-phosphate may be converted to glucose 1-phosphate, formed into a complex with uridine and added to a glycogen polymer by glycogen synthetase (uridine-diphosphoglucose-glycosyl transferase), or it may be oxidized to pyruvate via the glycolytic pathway and, depending on oxygen availability, be converted to lactate or completely oxidized in the mitochondria to CO_2 and water. An alternative route for glucose 6-phosphate is the pentose phosphate pathway whereby reduced NADPH is produced. Reduced NADP ($NADPH_2$) is utilized for fatty acid synthesis. About 30% of the glucose oxidized by the liver will pass through the pentose phosphate pathway. Finally, glucose 6-phosphate may be dephosphorylated by glucose 6-phosphatase to free glucose, which can diffuse back into the blood. The direction of glucose 6-phosphate metabolism is determined by the overall metabolic requirements of the organism at any moment and its dietary state. Metabolism is then altered accordingly by the endocrine environment.

The major lipids of importance in liver metabolism occur as triglycerides, long-chain fatty acids, short-chain fatty acids, cholesterol and cholesterol esters. For the moment, the first two groups will be considered. Fatty acids and glycerol are absorbed from the gut following enzymatic hydrolysis of triglycerides and are reconverted to triglycerides in the intestinal epithelial cells. Large droplets of triglycerides as well as some cholesterol and cholesterol esters are coated with lipoprotein and released into the intercellular spaces around the epithelial cell. These droplets or *chylomicrons* enter lacteals, the blind capillaries of the lymphatic system, and eventually enter the blood vascular system via the thoracic ducts. Chylomicrons are similar to the VLDLs produced by the liver except they do not contain cholesterol. The coating of chylomicrons and the triglycerides are hydrolyzed by lipoprotein lipase localized in the endothelial cells of capillaries in adipose and, to a lesser extent, muscle tissues, and released fatty acids and glycerol are directly absorbed by these tissues. The high local concentrations of these substances allow them to readily diffuse into adipose or muscle cells.

Nonesterified fatty acids and glycerol released from hydrolysis of triglycerides in the gut may diffuse directly into intestinal capillaries, bind to albumin and travel to the liver via the portal vein. Uptake of NEFA and glycerol is simply a function of concentration, and since intracellular NEFA and glycerol are essentially nonexistent because of their rapid metabolism, a considerable quantity may be removed from the blood with ease. Although 99.9% of the plasma NEFAs form a complex with albumin, the total capacity for NEFA transport is rather low. Consequently most long-chain fatty acids enter the system as triglycerides in chylomicrons.

Nonesterified fatty acids and glycerol accumulated by the liver cell may be used for energy through β-oxidation, for glucose synthesis (glycerol only) or for long chain fatty acid and triglyceride synthesis. Liver cells of mammals are capable of storing relatively small amounts of triglycerides.

Liver cells oxidize NEFAs to acetyl-CoA when conditions dictate mobilization of fatty acids and gluconeogenesis. Much of the acetyl-CoA is converted to ketone bodies as previously described. Since liver cells lack the necessary enzymes to metabolize ketone bodies further, they are released into the blood. Other tissues especially muscle and brain, remove these ketone bodies, reconvert them to acetyl-CoA and synthesize ATP. As in the case of NEFA oxidation, ketone bodies do not provide net sources of glucose because of the production of CO_2 from acetate in the TCA cycle.

If glycogen and glucose are limited as a consequence of lowered blood glucose, amino acids may be deaminated (transamination) to provide gluconeogenic substrates. Under conditions of glycogen depletion the liver must maintain sufficient circulating glucose through gluconeogenic processes to provide adequate energy sources for the brain and other obligate users of glucose (for example, red blood cells). Note that the cortical portion of the kidney is capable of carrying out gluconeogenesis as well as limited oxidation of glucose or storage as glycogen, and it aids the liver in maintaining sufficient levels of glucose during periods of food deprivation or low dietary intake of carbohydrates.

Excessive levels of amino acids stimulate gluconeogenesis in liver cells. If glycogen stores are adequate, the carbohydrate produced is converted to fatty acids and glycerol. Triglycerides are synthesized and exported as VLDLs destined for storage in adipose tissue.

Adipose Tissue

Adipose tissue functions primarily as a storage reservoir for respiratory fuel. Nonesterified fatty acids, long chain fatty acids and glycerol obtained from hydrolysis of chylomicrons or VLDL in the capillary beds of adipose tissue are readily accumulated by adipose cells, and the fatty acids are esterified with glycerol to triglycerides. Neutral fats are an excellent storage form because 1 g of fat when hydrolyzed produces nine times more energy per gram than does an equivalent weight of glucose stored as glycogen. Consequently much more stored fat fuel can be packed into a limited space. If migratory birds, such as the ruby-throated hummingbird that migrates hundreds of miles nonstop over ocean, were to rely solely on stored glycogen, they would be too heavy to fly!

Triglycerides can be hydrolyzed on demand by a lipase complex consisting of triglyceride lipase, diglyceride lipase and monoglyceride lipase to yield long-chain fatty acids and glycerol. Triglyceride lipase appears to be the rate-limiting enzyme in this pathway and the one that is regulated by hormones. Long-chain fatty acids are hydrolyzed to smaller fatty acids (NEFAs). Glycerol and NEFAs enter the blood where the latter form a complex with serum albumin and are transported to other tissues such as muscle or liver where they can be utilized as energy sources.

Glycolysis is regulated in adipose cells primarily by the availability of glucose. Therefore the rate of entrance of glucose into the cell determines the rate of glycolysis. Increased levels of glucose 6-phosphate augment formation of NADPH via the pentose phosphate pathway and synthesis of glycerol phosphate, both of which enhance lipogenesis. A decrease in glucose 6-phosphate similarly enhances lipolysis. This relationship between glycolysis and lipid metabolism is the basis for the *glucose-fatty acid cycle*, which theoretically can regulate fatty acid mobilization simply by the availability of glucose. When glucose becomes scarce this mechanism would favour conservation of available glucose arid would provide additional substrates for energy metabolism (NEFA) and gluconeogenesis (glycerol).

Muscle

Muscle cells are specialized for oxidizing glucose to produce the ATP needed for sustaining contraction. Red muscle is especially rich in mitochondria, TCA enzymes and electron transport molecules for complete oxidation of glucose to CO_2 and water. In contrast, white muscles (such as the pectoral muscles of domestic chickens) are specialized for anaerobic oxidation of glucose to lactate and have few mitochondria. The account here deals only with red muscle. Because of the high rate of glycolysis and aerobic oxidation of glucose in muscle as compared to liver, most of the detailed knowledge of these biochemical pathways has been generated from studies of muscle. Liver does not carry on as much oxidative phosphorylation as does muscle, and the components in this system are more difficult to study in liver because of the low rates of reaction.

Uptake and oxidative metabolism of glucose by muscle cells is very similar to events occurring in liver, with only a few important differences. Glucose is accumulated by facilitated transport, but the enzyme, hexokinase, responsible for phosphorylation of glucose has a lower affinity for substrate than liver glucokinase and is inhibited by accumulation of product. Only two routes are readily available to glucose 6-phosphate in muscle. Synthesis of glycogen is accomplished by the glycogen synthetase system, and the musculature may store more total glycogen than does the liver. For example, in humans the liver stores about 70 g of glucose as glycogen, whereas the musculature stores about 120 g. However, muscle glycogen is not available to provide glucose for export from the muscle cell because, unlike liver cells, muscle cells lack glucose 6-phosphatase and cannot hydrolyze glucose 6-phosphate to free glucose that could diffuse into the blood.

The second pathway for metabolism of glucose 6-phosphate in muscle is glycolysis. When muscle is performing only moderate work, the blood can supply glucose to maintain sufficient ATP levels. Greater work loads will draw upon glycogen reserves, and if demand exceeds the ability of blood to supply sufficient O_2, muscular activity may be maintained for a limited time by rapid anaerobic metabolism of glucose to lactate. The pentose phosphate pathway for glucose metabolism is not very important in muscle.

Muscle cells may also utilize NEFAs directly as an energy source and can utilize VLDL synthesized by the liver as a source of NEFAs because, like adipose tissue, some muscle capillary beds have cells that produce lipoprotein lipase. However, the extent of the contribution of exogenous NEFA to overall muscle metabolism has not been determined. Ketone bodies when present in the blood are also a ready source of energy for muscle cells, which can resynthesize acetyl-CoA and under aerobic conditions synthesize ATP. Muscle cells may synthesize and store triglycerides, which can also be hydrolyzed to provide fatty acids for energy production. The ability for muscle to metabolize endogenous or exogenous lipids for energy becomes extremely important under conditions of fasting.

Brain

The brain and rest of the nervous system have a high specific requirement for glucose as the principle energy source. Nervous tissue has lost the ability to store glycogen, does not store triglycerides and must therefore rely exclusively on blood for its source of glucose. Deprivation of glucose for only a matter of minutes can produce cell death and irreparable damage to the brain. Red blood cells and cells of the kidney medulla and the testes have similar high glucose requirements, but these cells utilize glucose only anaerobically, whereas brain also has a high O_2 requirement for glucose oxidation. Since muscle glycogen cannot be released it is the liver, and to a limited extent the kidney cortex, that must provide glucose for these tissues through glycogenolysis and gluconeogenesis. Under conditions of starvation, alternative energy sources provide some glucose, and other substrates must be utilized by the brain for energy.

Hormones Regulating Metabolism

Metabolism is influenced by many hormones that affect the availability of substrates and the activity of key enzymes in glycolysis, gluconeogenesis, glycogenolysis, lipogenesis and lipolysis. These hormones include *insulin* and *glucagon* from the endocrine pancreas, *epinephrine* from the adrenal medulla, *glucocorticoids* from the adrenal cortex, *growth hormone* (GH) and *β-lipotropin* (LPH) from the adenohypophysis, *thyroid hormones* and *gonadal steroids*. In addition, the *cyclic nucleotides* (cAMP, cGMP) or the *prostaglandins* or both may be important mediators of the action of one or more of the above hormones. Insulin-like growth factors also influence metabolism. Metabolism may be affected by drugs such as caffeine that alter the activity of cyclic nucleotide-degrading enzymes (phosphodiesterases).

These metabolic hormones may be classified according to their general effects on metabolism of carbohydrates, fats or protein. Because of the existence of such interrelationships between these classes of compounds, many metabolic hormones influence all of them. Hormones may be considered glycolytic (insulin), glycogenolytic (epinephrine, glucagon, thyroid hormones) or gluconeogenic (epinephrine, GH, glucocorticoids). They may be lipogenic (insulin) or lipolytic (epinephrine, thyroid hormones, GH, LPH) with respect to mobilization of fatty acids. Some of these hormones are protein anabolic, that is, they increase the general synthesis of proteins and favour nitrogen retention (insulin, GH, thyroid hormones, androgens) or protein catabolic (glucocorticoids, thyroid hormones). Increased levels of circulating glucose (hyperglycemia) are caused by epinephrine, glucagon, thyroid hormones, GH, glucocorticoids and estrogens, whereas only insulin is hypoglycemic.

The following accounts of metabolism under differing physiological conditions emphasize the importance of insulin, glucagon, epinephrine, GH, and glucocorticoids as the major metabolic regulators, with glucocorticoids and thyroid hormones exhibiting permissive actions with respect to GH. The extent of involvement by LPH is not yet clear. The hormones that predominate under three conditions will be examined: regulation following digestion of a meal (postabsorptive), responses to acute stress and responses to chronic stress (starvation).

Endocrine Regulation Following Feeding

Insulin is the major hormone responsible for regulating carbohydrate, amino acid and lipid levels in omnivores and carnivores in the immediate postabsorptive state. Insulin increases the rate of transport of glucose into liver, muscle and adipose cells. In liver, this stimulates glycogenesis and production of VLDLs and LDLs. Glucose entry into muscle cells stimulates glycogenesis and glycolysis. This increase observed in muscle glycolysis appears to be due simply to an increase in substrates. In adipose tissues, glucose entry favours fatty acid synthesis and esterification, mainly by substrate effects.

Entry of amino acids into muscles stimulated by insulin enhances protein synthesis. In the liver and to some extent the kidney cortex, excess amino acids are deaminated (transaminated), and if sufficient glucose and glycogen are present, these molecules will contribute to lipogenesis for export as lipids to adipose tissue for storage. Insulin depresses gluconeogenesis in liver cells by favouring incorporation of amino acids into protein and also in adipose cells, although the latter have not been extensively studied.

In addition to effects of insulin on the cell membrane and transport of glucose and amino acids, insulin alters lipid and carbohydrate metabolism through effects on enzymes. In muscle and probably in liver, insulin interferes with the enzyme protein kinase by increasing the proportion of the enzyme that is inactive. Protein kinase is responsible for two important enzymatic interconversions. The first involves conversion of inactive phosphorylase-b-kinase to an active form, which in turn catalyzes phosphorylation of another enzyme, phosphorylase -b, to an active form, phosphorylase-a. Phosphorylase-a is responsible for cleaving a glucose 1-phosphate residue from glycogen. Conversion of glucose 1-phosphate to glucose 6-phosphate is determined by substrate levels. At the same time, protein kinase catalyzes inactivation of glycogen synthetase also through phosphorylation. Thus, phosphorylation via protein kinase activates glycogenolysis (phosphorylase-a) and inhibits glycogenesis (glycogen synthetase), resulting in an increase in intracellular glucose 6-phosphate. Insulin favours glycogen synthesis by inhibiting protein kinase activity. This effect may be caused by an insulin-induced reduction in cAMP related to an insulin-induced increase in cGMP.

Insulin also increases the amount of glucokinase in liver cells and reduces the amount of glucose 6-phosphatase. These changes seem to reflect alterations in rates of synthesis of these enzymes and would increase glycolysis. There are no apparent effects on these enzyme levels in either muscle or adipose tissue.

NEFA synthesis and triglyceride synthesis are increased in liver by insulin. In adipose tissue, insulin increases synthesis of lipoprotein lipase, which enhances lipogenesis. Insulin also stimulates triglyceride synthesis in adipose tissue from either glucose or acetyl-CoA, but the mechanism of this effect is not fully understood. Lipolysis is inhibited by insulin in adipose cells by a reduction in activity of triglyceride lipase that is correlated with a reduction in cAMP levels. Thus, insulin in adipose tissue is both lipogenic, largely through effects on substrate availability, and antilipolytic, through an effect on cAMP levels. Although the effects of insulin on lipid metabolism in liver and adipose tissue are marked, there seems to be no influence of insulin on lipid metabolism in muscle.

Immediately following a high protein meal, then, insulin may be considered the most important single hormone since it directs excess fuels obtained from the diet into storage. Somewhat later, GH and insulin work cooperatively in that GH stimulates some lipolysis, which provides utilization of fatty acids for meeting energy requirements of protein anabolic processes that are being stimulated by GH. Such cooperative action also conserves glucose for those tissues that have specific requirements for glucose (for example, the brain) while encouraging others (muscle) to use an alternative fuel.

Herbivore metabolism differs markedly from that of carnivores. Most studies have been done with ruminants. Digestion in these large herbivores is accomplished by bacterial flora that reside in a special region of the gut known as the rumen. The bacteria digest cellulose to release glucose. This glucose is utilized by the bacteria and is not available for absorption into the ruminant. Instead, the ruminant absorbs fatty acids and amino acids resulting from metabolism and death of bacteria. Glucagon is the main hormone released after feeding, and it stimulates gluconeogenesis from amino acids. Studies of cattle and sheep indicate that insulin increases incorporation of lipids into adipose tissue. Growth hormone directs energy to skeletal muscle. During periods between feedings, the glucocorticoids are responsible for maintaining gluconeogenesis.

Effects of Acute Stress on Metabolism

Acute stress such as the "flight-or-fight" response and exercise produces profound alterations in energy metabolism that are mediated primarily by epinephrine released from the adrenal medulla and to some extent by glucagon from the endocrine pancreas. Hyperglycemia can occur following release of epinephrine due to a direct inhibition of insulin release.

In muscle, epinephrine activates adenyl cyclase and increases cAMP with resultant activation of protein kinase, as discussed previously. Increased activity of protein kinase accelerates glycogenolysis and inhibits glycogen synthesis in muscle even when insulin levels are high during the immediate postabsorptive state. Since muscle lacks glucose 6-phosphatase, the released glucose is utilized within the muscle cell for oxidative phosphorylation as is glucose taken up from the blood. Although epinephrine has been shown experimentally to activate glycogenolysis in liver, large (pharmacological) doses are required, and it is probably glucagon that is responsible for effects on glycogenolysis observed under physiological conditions. Reduced blood sugar levels due to increased glucose utilization would be sufficient to induce glucagon release, which through the same cAMP mechanism would stimulate glycogenolysis in liver.

Epinephrine also stimulates cAMP production in adipose tissue where it increases activity of triglyceride lipase, resulting in an increase in blood NEFAs and glycerol. A similar effect may occur in those muscles where triglyceride stores have been produced. The release of fatty acids and their use for fuel during prolonged exercise such as marathon running protects liver glycogen stores from metabolism by muscle and makes them available for tissues with high glucose requirements such as brain.

The increase in plasma NEFAs and their uptake by liver tends to inhibit glycolytic enzymes and increase acetyl-CoA for ATP synthesis or ketone body formation. Accumulation of acetyl-CoA favours conversion of pyruvate obtained from lactate to oxaloacetate and resynthesis of glucose via the gluconeogenic pathway. Thus, increases in fatty acids can influence glucose availability in association with the glucose-fatty acid cycle.

Effects of Chronic Stress (Starvation) on Metabolism

During fasting or early starvation in humans there is an elevation in GH levels as well as in glucagon levels. The liver is capable of storing glucose in the form of glycogen sufficient for only about 12 to 24 hours of fasting, during which time all of the stored liver glucose is depleted. Fatty acid mobilization and oxidation become the major source of energy for ATP synthesis in muscle, and gluconeogenesis from amino acids in the liver and kidney cortex becomes the primary source for blood glucose. Growth hormone and possibly LPH stimulate lipolysis in adipose tissue and provide the required fatty acids. Glycerol also produced from triglyceride hydrolysis can be used as a gluconeogenic precursor. The action of GH on adipose tissue requires the presence of glucocorticoids. Furthermore, glucocorticoids inhibit peripheral utilization of amino acids for protein synthesis and increase levels of transaminases in liver, which are necessary for converting amino acids to ketoacids for gluconeogenesis. The latter effect may be a secondary action of glucocorticoids but nevertheless contributes significantly to the altered metabolism of starvation.

Glucagon normally stimulates glycogenolysis in liver and enhances entry of amino acids into the gluconeogenic pathway. The resultant increase in blood glucose under nonemergency conditions enhances glucose uptake by adipose tissue and stimulates lipogenesis. However, in the fasting animal, glucose levels cannot be maximally elevated by glucagon because of lack of stored glycogen reserves. Instead, glucagon stimulates lipolysis by a cAMP-dependent activation of triglyceride lipase. These effects of glucagon and GH on adipose tissue become even more important during later stages of starvation when there is a greater emphasis on consumption of lipid fuels than on using amino acids via gluconeogenesis.

Oxidation of fatty acids by liver produces large amounts of acetyl-CoA, of which a considerable proportion under starvation conditions is converted to ketone bodies and secreted into the blood. Ketone bodies are removed from the blood by muscle, which reconverts them to acetyl-CoA for ATP synthesis. Thus an increase in ketone body production reduces glucose requirements by the musculature and the need for catabolism of amino acids. Ketone bodies, like fatty acids, depress glycolysis and favour their own utilization as energy sources. The brain, which normally has a high requirement for glucose (almost 100% of its energy demands are met by blood glucose), switches to predominant use of ketone bodies and drastically reduces its requirement for glucose. This lessened demand for glucose further reduces reliance on protein and amino acids for energy while at the same time encouraging use of triglyceride stores. In humans this switch by the brain from glucose to ketone bodies occurs after about 3 to 10 days of starvation, but the mechanism responsible for this switch is not known. If this switch did not occur, however, protein would be preferentially degraded over fat for energy, resulting in serious protein deficiencies, and would reduce significantly the survival time during total starvation. It has been calculated that if this change from amino acid emphasis and attendant nitrogen loss via urea production and excretion did not occur, death would ensue within about 21 days. However, it is well known that obese people tolerate starvation for as long as 240 days without ill effects largely because of this reduction in reliance on glucose by the nervous system and other tissues. Eventually the fat reserves will be depleted, and the body must again rely on proteins and gluconeogenesis, with rapid appearance of protein deficiencies that, if not alleviated, will lead to death. Under conditions of semistarvation (insufficient caloric intake such that some body reserves must be consumed daily to meet energy requirements of the organism), the switch over to ketone body utilization and fat depletion does not occur, and protein deficiencies become prevalent very early. Under total starvation protein deficiencies do not appear until much later, when lipid reserves have been depleted.

Role of Protein Anabolic Steroids in Metabolism

Many hormones have been shown to produce nitrogen retention (reduction in nitrogen excretion as urea due mainly to increased protein synthesis) including GH, thyroid hormones, estrogens and insulin.

Kochakian and co-workers in the mid 1930s demonstrated that androgens also produced nitrogen retention, but the mechanism of their action was very different from the above hormones. Growth hormone, for example, stimulates new protein synthesis primarily in tissues other than muscle, whereas androgens produce their most marked effects in the musculature. Estrogens produce their protein anabolic effects only on selected tissues (uterus, mammary gland, female genital tract, skin, skeleton) and have rather little effect on overall nitrogen balance. Progesterone, which is clearly protein anabolic with respect to certain target tissues such as the uterus, actually produces a negative effect on nitrogen balance (increases total nitrogen excretion) and cannot be considered protein anabolic with respect to overall metabolism. Thyroid hormones are protein anabolic in thyroidectomized but not hypophysectomized animals, suggesting that this effect is mediated through GH release.

The first claim for rejuvenating (protein anabolic) effects of androgens stems from the experiments of Brown-Sequard in 1889, who reported effects of extracts prepared from animal testes on himself. Since these unsubstantiated claims, testosterone and other androgenic steroids have been shown to cause stimulation of protein synthesis, which is supported by glycogen reserves and influenced by availability of protein in the diet. The observation that it is not possible to dissociate the protein anabolic effect (nitrogen retention) from the androgenic effect has resulted in formulation of the descriptive title *anabolic-androgenic steroid* for such compounds. Androgens do not apparently influence amino acid transport but increase activity of amino acid activating enzymes. The effect of anabolic-androgenic steroids wears off with continued treatment, however, and nitrogen metabolism returns to normal.

The potential effect of anabolic-androgenic steroids on strength and endurance has resulted in their use for athletic training programs. Few adequately controlled and properly designed studies have been published. Results are divided between claims of improved performance and no effect. Treatment with synthetic testosterone analogues depresses secretion of gonadotropins and testosterone and decreases spermatogenesis. Furthermore, liver disfunction was reported in 80% of persons examined who had taken C_{17}-alkylated derivatives of testosterone. There are numerous reports of hepatitis, liver failure and fatal liver cancer related to use of protein anabolic-androgenic steroids. These physiological effects and their implications on the health of the user must be weighed against the short-term questionable benefits.

6

ADRENAL MEDULLA

The adrenal medulla was first distinguished from the adrenal cortex at the beginning of the 19th century. Its major secretory product, *epinephrine*, was isolated, purified, and synthesized a century later. *Norepinephrine* was synthesized in 1904, but not until 1946 was it recognized as a secretory product of the adrenal medulla and as the major neurotransmitter of the postganglionic sympathetic nerves. The adrenal medulla, a highly specialized part of the sympathetic nervous system, functions under stress or whenever marked deviations from normal homeostasis occur—in contrast to the rest of the sympathetic nervous system, which is involved in the minute-to-minute fine regulation of most physiologic processes.

ANATOMY

Embryology

The sympathetic nervous system arises in the fetus from the primitive cells of the neural crest (sympathogonia). At about the fifth week of gestation, these cells migrate from the primitive spinal ganglia in the thoracic region to form the sympathetic chain posterior to the dorsal aorta. They then begin to migrate anteriorly to form the remaining ganglia.

At 6 weeks, groups of these primitive cells migrate along the central vein and enter the fetal adrenal cortex to form the adrenal medulla, which is detectable by the eighth week. The adrenal medulla at this time is composed of sympathogonia and pheochromoblasts, which then mature into pheochromocytes. The cells appear in rosettelike structures, with the more primitive cells occupying a central position. Storage granules can be found in these cells by electron microscopy at

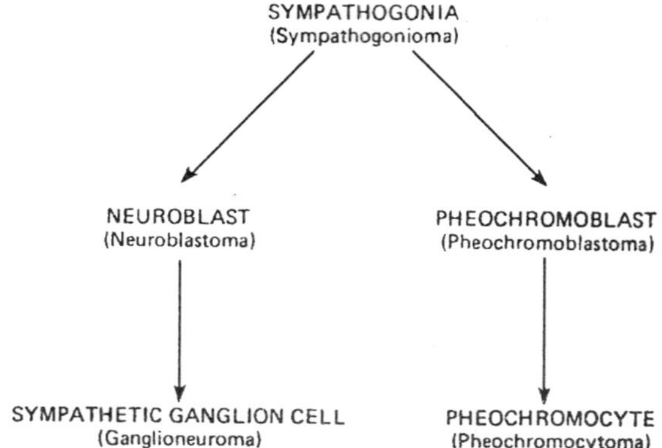

Fig. 6.1. The embryonic development of adrenergic cells and tumors that develop from them. Sympathogonia are primitive cells derived from the neural crest.

12 weeks. Pheochromoblasts and pheochromocytes also collect on both sides of the aorta to form the paraganglia. The principal collection of these cells is found at the level of the inferior mesenteric artery. They fuse anteriorly to form the organ of Zuckerkandl, which is quite prominent in fetal life. This organ is thought to be a major source of catecholamines during the First year of life, after which it begins to atrophy. Pheochromocytes (chromaffin cells) also are found scattered throughout the abdominal sympathetic plexuses as well as in other parts of the sympathetic nervous system.

Gross Structure

The anatomic relationships between the adrenal medulla and the adrenal cortex are different in different species. These organs are completely separate structures in the shark. They remain separate but in close contact in amphibians, and there is some intermingling in birds. In mammals, the medulla is surrounded by the adrenal cortex. In humans, the adrenal medulla occupies a central position in the widest part of the gland, with only small portions extending into the narrower parts. It constitutes approximately one-tenth of the weight of the gland, although the proportions vary from individual to individual. There is no clear demarcation between cortex and medulla. The central vein is usually surrounded by a cuff of adrenal cortical cells, and there may be islands of cortex elsewhere in the medulla.

Microscopic Structure

The *chromaffin cells*, or *pheochromocytes*, of the adrenal medulla are large ovoid columnar cells arranged in clumps or cords around blood vessels. They derive their name from the observation that their granules turn brown (*pheo-*) when stained with chromic acid. The colour is due to the oxidation of epinephrine and norepinephrine to melanin. These cells have large nuclei and a well-developed Golgi apparatus. They contain large numbers of vesicles or granules containing catecholamines. Vesicles containing norepinephrine are darker than those containing epinephrine.

The pheochromocytes may be arranged in nests, alveoli, or cords and are surrounded by a rich network of capillaries and sinusoids. The adrenal medulla also contains some sympathetic ganglion cells, singly or in groups. Ganglion cells are also found in association with the viscera, the carotid body, the glomus jugulare, and the cervical and thoracic ganglia.

Nerve Supply

The cells of the adrenal medulla are innervated by preganglionic fibers of the sympathetic nervous system, which release acetylcholine at the synapses. These fibers arise most commonly from a plexus in the capsule of the posterior surface of the gland and enter the adrenal glands in bundles of 30-50 fibers without synapsing. They follow the course of the blood vessels into the medulla without branching into the adrenal cortex. Some reach the wall of the central vein, where they synapse with small autonomic ganglia. However, most fibers end in relationship to the pheochromocytes.

Blood Supply

The human adrenal gland derives blood from the superior, middle, and inferior adrenal branches of the inferior phrenic artery directly from the aorta, and from the renal arteries. Upon reaching the adrenal gland, these arteries branch to form a plexus under the capsule supplying the adrenal cortex. A few of these vessels, however, penetrate the cortex, passing directly to the medulla. The medulla is also nourished by branches of the arteries supplying the central vein and cuff of cortical tissue around the central vein. Capillary loops passing from the subcapsular plexus of the cortex also supply blood as they drain into the central vein. It would appear, then, that most of the blood supply to the medullary cells is via a portal vascular system arising from the capillaries in the cortex. There is also a capillary network of lymphatics that drain into a plexus around the central vein.

In mammals, the enzyme that catalyzes the conversion of norepinephrine to epinephrine (phenylethanolamine-N-methyltransferase, PNMT) is induced by adrenocortical steroids. The chromaffin cells containing epinephrine therefore receive most of their blood supply from the capillaries draining the cortical cells, whereas cells containing predominantly norepinephrine are supplied by the arteries that directly supply the medulla.

On the right side, the central vein is short and drains directly into the vena cava, although some branches go to the surface and reach the azygos system. On the left, the vein is somewhat longer and drains into the renal vein.

Hormones of the Adrenal Medulla

Catecholamines

Biosynthesis

Catecholamines are widely distributed in plants and animals. In mammals, *epinephrine* is synthesized mainly in the adrenal medulla, whereas *norepinephrine* is found not only in the adrenal medulla but also in the central nervous system and in the peripheral sympathetic nerves. *Dopamine*, the precursor of norepinephrine, is found in the adrenal medulla and in noradrenergic neurons. It is present in high concentrations in the brain, in specialized interneurons in the sympathetic ganglia, and in the carotid body, where it serves as a neurotransmitter. Dopamine is also found in specialized mast cells and in enterochromaffin cells.

The proportions of epinephrine and norepinephrine found in the adrenal medulla vary with the species. In humans, the adrenal contains 15-20% norepinephrine.

Tyrosine as precursor

The catecholamines are synthesized from *tyrosine*, which may be derived from ingested food or synthesized from phenylalanine in the liver. Tyrosine circulates at a concentration of 1-1.5 mg/dL of blood. It enters neurons and chromaffin cells by an active transport mechanism and is converted to *dihydroxyphenylalanine* (*dopa*). The reaction is catalyzed by *tyrosine hydroxylase*, which is transported via axonal flow to the nerve terminal. Tyrosine hydroxylase activity may be inhibited by a variety of compounds. Alpha-methylmetatyrosine is effective and is sometimes used in the therapy of malignant pheochromocytomas. Substances that chelate iron or compete for the pteridine cofactor also inhibit the enzyme but are not useful clinically. This enzyme is a reliable marker of intact sympathetic nerve tissue and neurotransmitter synthesis.

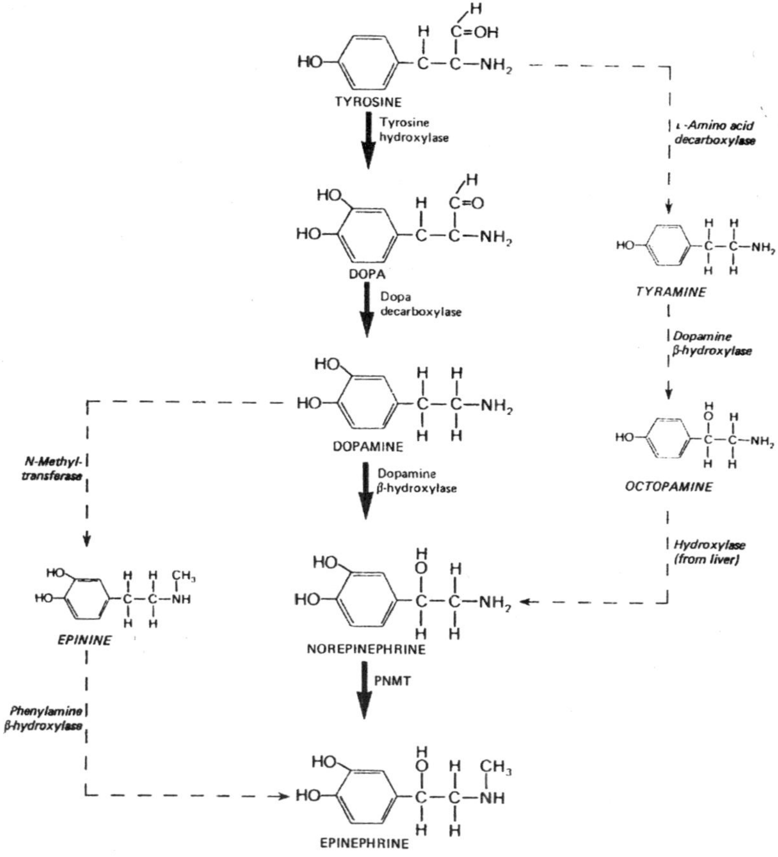

Fig. 6.2. Biosynthesis of catecholamines.

Conversion of dopa to dopamine

Dopa is converted to dopamine by the enzyme aromatic L-amino acid decarboxylase (dopa decarboxylase). This enzyme is found in all tissues, with the highest concentrations in liver, kidney, brain, and vas deferens. The various enzymes have different substrate specificities depending upon the tissue source. Competitive inhibitors of dopa decarboxylase such as methyldopa are converted to substances (an example is α-methylnorepinephrine) that are then stored in granules in the nerve cell and released in place of norepinephrine. These products (false transmitters) were thought to mediate the antihypertensive action of drugs at peripheral sympathetic synapses but are now believed to stimulate the a receptors of the inhibitory corticobulbar system, reducing sympathetic discharge peripherally.

The conversion of dopamine to norepinephrine is catalyzed by dopamine β-hydroxylase, a mixed function oxidase requiring oxygen and an external electron donor. The enzyme does not occur in tissues outside of the neuron. It has activities consistent with the 2 alternative pathways for catecholamine synthesis. Neither of these alternative path ways is thought to be physiologically important.

Part of the biologic specificity of dopamine β-hydroxylase may result from its compartmentalization. Newly synthesized dopamine β-hydroxylase is incorporated directly into storage vesicles that take up, synthesize, and store the catecholamines. The membranes of these vesicles contain dopamine β-hydroxylase, ATPase, cytochrome P-561, and cytochrome P-561:NADH reductase. Dopamine β-hydroxylase is also found within the granule and is released with norepinephrine during secretion. Inhibitors of the enzyme, such as disulfuramic and picolinic acid, have no clinical importance.

Conversion of norepinephrine to epinephrine

PNMT catalyzes the N-methylation of norepinephrine to epinephrine, using S-adenosylmethionine as methyl donor. It is found only in the adrenal medulla and in a few neurons in the central nervous system. The enzyme is found in the cytosol. Norepinephrine leaves the granule and after methylation reenters different granules. It is induced by the high levels of glucocorticoids (100 times the systemic concentration) found in the adrenal medulla. Inhibitors of this enzyme can reduce blood pressure in hypertensive rats. The conversion of dopamine to *epinine* is catalyzed by a nonspecific N-methyltransferase.

Catecholamine biosynthesis is coupled to secretion, so that the stores of norepinephrine at the nerve endings remain relatively unchanged even in the presence of marked nerve activity. In the adrenal medulla, it is possible to deplete stores with prolonged hypoglycemia, but no significant change occurs with splanchnic nerve stimulation. Biosynthesis appears to be increased during nerve stimulation by activation of tyrosine hydroxylase. Prolonged stimulation leads to the induction of increased amounts of this enzyme.

Storage

The catecholamines are found in the adrenal medulla and various sympathetically innervated organs, and their concentration reflects the density of sympathetic neurons. The adrenal medulla contains about 0.5 mg/g; the spleen, vas deferens, brain, spinal cord, and heart contain 1-5 μg/g; liver, gut, and skeletal muscle contain 0.1-0.5 μg/g The catecholamines are stored in electron-dense granules approximately 1 μm in diameter that contain catecholamines and lipids, ATP in a 4:1 molar ratio, calcium, magnesium, and water-soluble proteins called *chromogranins*. The ratio of catecholamines to ATP is much higher in granules isolated from pheochromocytomas. The interior surface of the membrane contains dopamine β-hydroxylase and ATPase. The Mg^{2+} -dependent ATPase facilitates the uptake and inhibits the release of catecholamines by the granules. This activity is inhibited by reserpine.

Secretion

Adrenal medullary catecholamine secretion is increased by exercise, angina pectoris, myocardial infarction, hemorrhage, ether anesthesia, surgery, severe hypoglycemia, anoxia and asphyxia, and many other stressful stimuli. The rate of secretion of epinephrine increases more than that of norepinephrine during hypoglycemia and most other stimuli. However, anoxia and asphyxia selectively increase adrenal medullary release of norepinephrine.

Secretion of the adrenal medullary hormones is mediated by the release of acetylcholine from the terminals of preganglionic fibers. The resulting depolarization of the axonal membrane triggers an influx

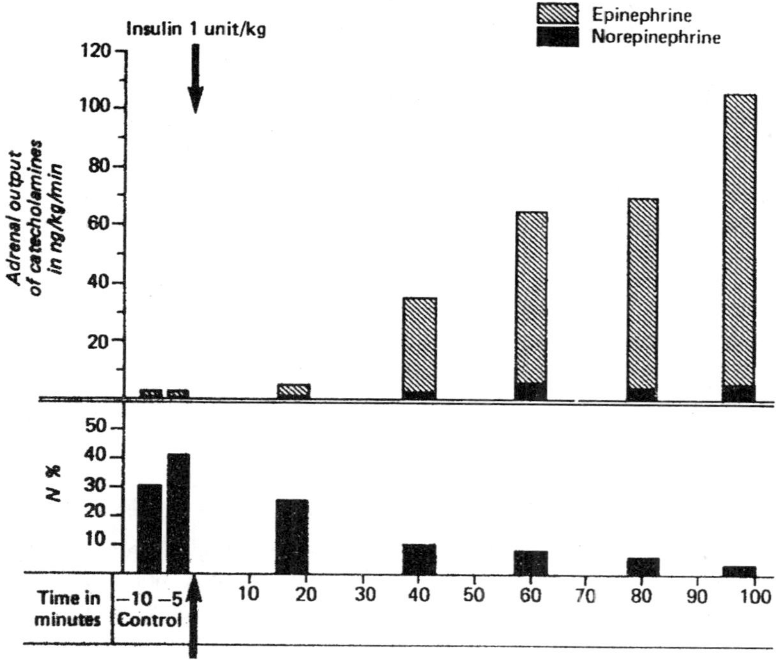

Fig. 6.3. Rate of secretion of amines from one adrenal following injection of insulin.

of calcium ion. The contents of the storage vesicles, including the chromogranins and soluble dopamine β-hydroxylase are released by exocytosis by the calcium ion. Membrane-bound dopamine β-hydroxylase is not released. *Tyramine*, however, releases norepinephrine primarily from the free store in the cytosol. Cocaine and monoamine oxidase inhibitors inhibit the effect of tyramine but do not affect the release of catecholamines by nervous stimulation. The rate of release in response to nerve stimulation is increased or decreased by a wide variety of neurotransmitters acting at specific receptors on the presynaptic neuron. The accumulation of excess catecholamines that are not in the storage granules is prevented by the presence of intraneuronal monoamine oxidase. Norepinephrine has an important role in modulating its own release by activating the α receptors on the presynaptic membrane. Alpha receptor antagonists inhibit this reaction. Conversely, presynaptic β receptors enhance norepinephrine release, whereas β receptor blockers increase it.

Transport

When released into the circulation, the amines are bound to albumin or a closely associated protein with low affinity and high capacity ($K_d \cong 10^{-7}$).

Metabolism and inactivation

The actions of catecholamines are terminated rapidly by several mechanisms. These include reuptake by the sympathetic nerve ending, metabolism by the enzymes catechol-O-methyltransferase (COMT) and monoamine oxidase, conjugation with sulfate ion, and direct excretion by the kidney. Estimates of the metabolic clearance rates for epinephrine and norepinephrine are 4000-5000 L/d.

Uptake 1

A large proportion (85-90%) of the amines released at the synapse are taken up locally by the nerve endings from which they are released (uptake 1). Circulating amines can also be taken up by this mechanism, which, however, plays a less important role in the inactivation of circulating catecholamines. The axonal membrane uptake process is energy-requiring, saturable, stereoselective, and sodium-dependent. It is blocked by cocaine, other sympathomimetic amines such as metaraminol and amphetamine, and agents such as tricyclic antidepressants and phenothiazines. Neither calcium nor magnesium ion has any effect on uptake 1. There is also an uptake process for dopamine in neurons of the central nervous system, but this process is

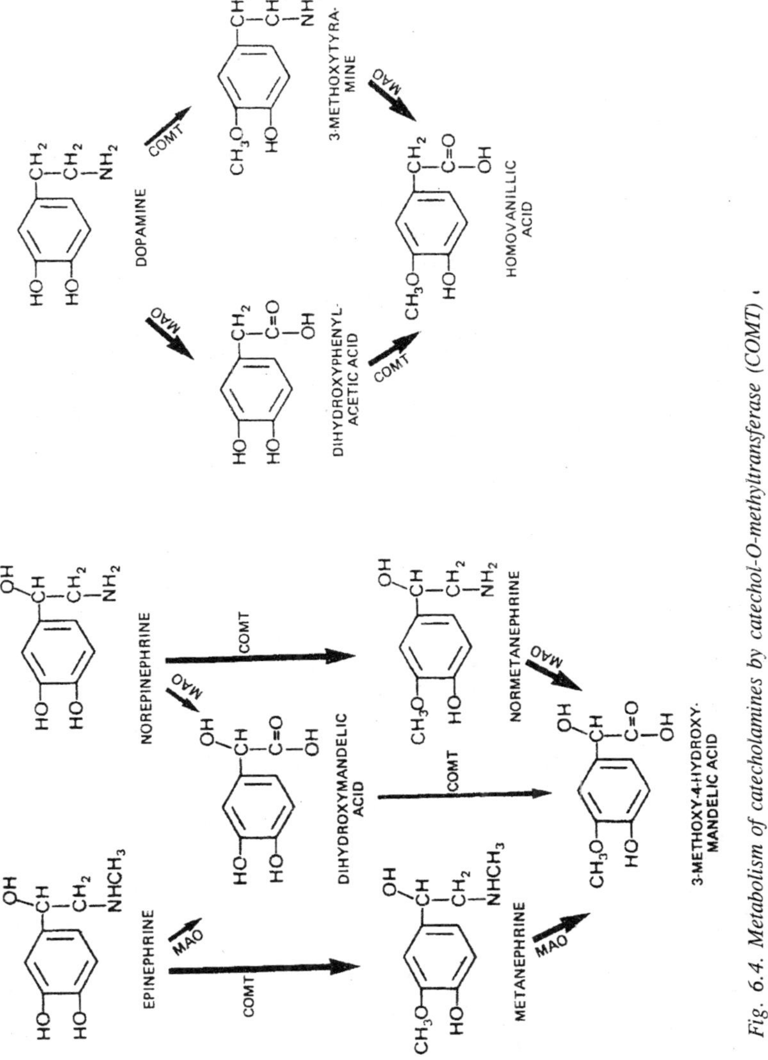

Fig. 6.4. Metabolism of catecholamines by catechol-O-methyltransferase (COMT).

not inhibited by tricyclic antidepressants. After being taken up by the neuron, the amines are reutilized or deaminated by monoamine oxidase and the metabolites are released. The intraneuronal monoamine oxidase is a flavoprotein and is localized to the outer mitochondrial membrane. Its substrate specificity favours norepinephrine, which it oxidizes to dihydroxyphenylglycol (DOPEG) or dihydroxymandelic acid (DOMA). The major functions of monoamine oxidase are (1) to regulate dopamine

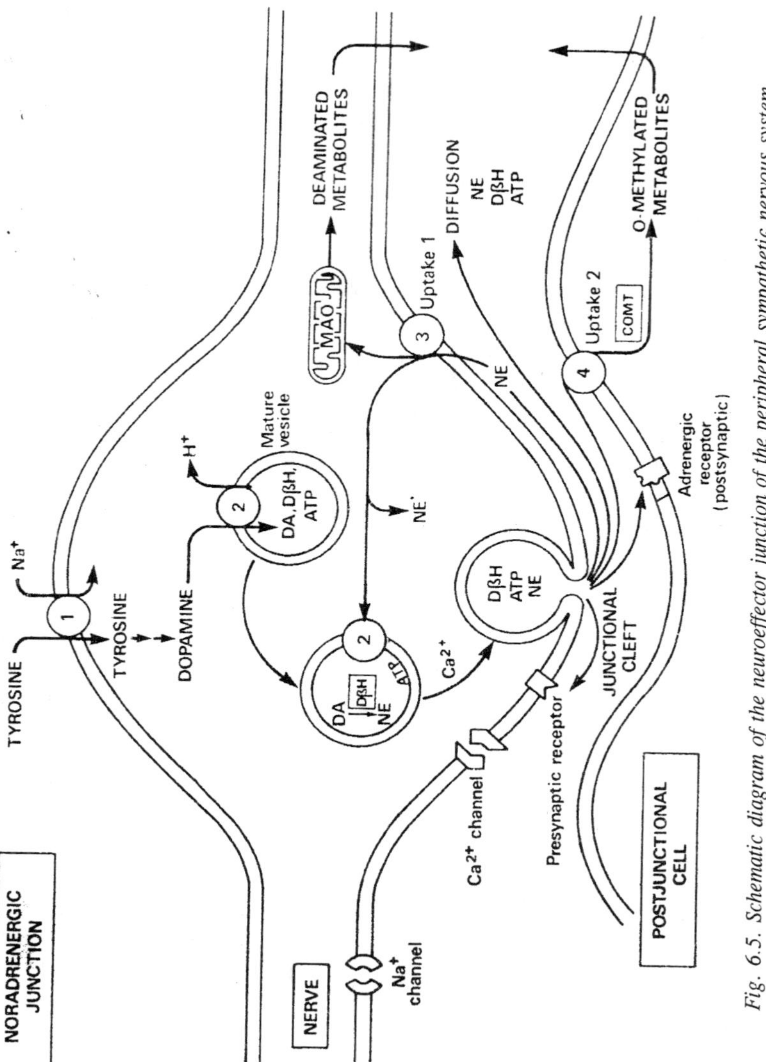

Fig. 6.5. Schematic diagram of the neuroeffector junction of the peripheral sympathetic nervous system.

and norepinephrine content of the neurons; (2) to destroy ingested amines: and (3) to metabolize the circulating catecholamines and their O-methyl metabolites. Two forms of monoamine oxidase with distinctive pharmacologic properties have been identified. Depressed patients have higher platelet monoamine oxidase type B contents than do normal controls. Progesterone increases the level of monoamine oxidase in women, whereas estrogens inhibit the enzyme.

Uptake 2

Extraneuronal tissues also take up catecholamines. This process, uptake 2, is saturable, is not specific for catecholamines, and is inhibited by various steroids, phenoxybenzamine, and normetanephrine but not by cocaine and other drugs that inhibit neuronal uptake.

At uptake 2, the catecholamines are metabolized to their O-methyl derivatives by COMT. This enzyme is found mainly in the soluble fraction of tissue homogenates, with the highest levels in the liver and kidney. S-Adenosylmethionine is the methyl donor, and divalent ions are required. It is predominantly an extraneuronal enzyme and acts on circulating catecholamines as well as on the locally released norepinephrine. It is the most important enzyme in the metabolism of circulating amines. Approximately 70% of circulating epinephrine in humans is methoxylated and about 24% is also deaminated. These enzymes also metabolize dopamine to homovanillic acid. Although normetanephrine is fairly active at the a receptor of the nictitating membrane, the metabolites of the catecholamines are generally devoid of biologic or physiologic activity.

Conjugation

The phenolic hydroxyl group of the catecholamines and their metabolites may be conjugated with sulfate or glucuronide. Liver and gut appear to be important sites of this reaction, and recent studies suggest that circulating red blood cells are a significant site of sulfation in humans.

Connective tissue binds catecholamines by a process that is inhibited by oxytetracycline. The significance of this binding is unknown.

Receptors

The catecholamines exert their physiologic effects by binding to receptor molecules on the surfaces of target cells. Early studies of sympathetic activation suggested that there were 2 classes of responses designated as inhibitory or excitatory. Cannon and his colleagues in the 1930s postulated that this difference was produced by the release of different neurotransmitters: sympathin I and sympathin E. The discovery by von Euler and others that norepinephrine was the neurotransmitter released by peripheral sympathetic nerve endings and was responsible for both types of responses led to the proposal by Ahlquist in 1948 that there were 2 types of receptors that he designated α and β, based on the relative potencies of a series of adrenergic agonists. The subsequent development of relatively specific antagonists confirmed this hypothesis and allowed the development of binding assays for the α, β, and dopaminergic receptors. Subsequent pharmacologic and biochemical studies indicate that α, β, and dopaminergic receptors can be further divided into subtypes.

Types of receptors

Two subtypes of this receptor have been identified and designated α_1 and α_2. It was initially thought that the α_2 receptor was limited to the presynaptic nerve ending and, when activated, served to inhibit the release of norepinephrine. However, these receptors have been found in platelets and postsynaptically in combination with α_1 receptors in adipose tissue and smooth muscle. The physiologic processes mediated by α receptors are dependent on the increased availability of intracellular calcium ion. However, the events leading to this increase have not been established. There is evidence that the α_2 receptor decreases the activity of adenylate cyclase, resulting in a reduction of cAMP formation. The activation of α_1 receptors enhances the breakdown of phosphotidylinositol and may act by altering the phospholipid structure of the cell membrane.

Two subtypes of the β receptor, β_1 and β_2, have been identified. Beta$_1$ receptor appears to mediate the inotropic and chronotropic effects on cardiac muscle, whereas β_2 receptors mediate smooth muscle relaxation in the bronchi and vascular smooth muscle as well as the uterus and adipose tissue. Norepinephrine is a powerful agonist at the β_1 receptor but comparatively weak at the β_2 receptor, compared to epinephrine or isoproterenol. Relatively selective β agonists have also been developed.

Dopaminergic receptors are found in the central nervous system, pituitary, heart, renal and mesenteric vascular beds, and other sites. Two subtypes of the dopaminergic receptor, D_1 and D_2, have been identified. The effects of the D_1 receptor are mediated by the adenylate cyclase system and are found postsynaptically in the brain. Those in the pituitary are D_2 receptors.

Receptor as site of regulation

The receptor also serves as a site of regulation of adrenergic activity. As noted above, presynaptically, norepinephrine released during nerve stimulation binds to a receptors and reduces the amount of norepinephrine released. Nerve endings have also been found to have receptors for many other agents presynaptically.

The number of receptors on the effector cell surface can be reduced by binding of agonist to receptor (antagonists do not have the same effect). This reduction is called "down-regulation." Thyroid hormone, on the other hand, has been shown to increase the number of β receptors in the myocardium. Estrogen, which increases the number of α receptors in the myometrium, increases the affinity of some vascular a receptors for norepinephrine.

Although the receptors serve as a site of fine regulation, the major physiologic control is exerted by alterations in activity of the nervous system. The discovery of cAMP and elucidation of its role in controlling physiologic processes—and the finding that most cells in the body have been found to have adrenergic receptors—have led to an appreciation of the important regulatory role of the peripheral sympathetic nervous system. In contrast, the effects of the adrenal medulla are mediated via the circulating amines and, therefore, are much more generalized in nature. Furthermore, adrenal medullary secretion increases significantly only in the presence of stress or marked deviation from homeostatic or resting conditions. For example, a minor reduction in the available glucose leads to sympathetic activation of fat mobilization from adipose tissue, whereas the adrenal medulla may not release large amounts of epinephrine until the blood sugar falls to 40 mg/dL in normal individuals.

Physiologic effects

Cardiovascular effects

Catecholamines increase the rate and frequency of contraction and increase the irritability of the myocardium by activating myocardial β_i receptors. The contractile effects of the catecholamines on vascular smooth muscle are mediated via α receptors. Although β receptors are present and cause dilatation, other mechanisms of vascular dilatation are probably more important. The release or injection of catecholamines can therefore be expected to increase heart rate and cardiac output and cause peripheral vasoconstriction—all leading to an increase in blood pressure. These events are modulated by reflex mechanisms, so that, as the blood pressure increases, reflex stimulation may slow the heart rate and tend to reduce cardiac output. Although norepinephrine in the usual doses will have these effects, the effect of epinephrine may vary depending on the smooth muscle tone of the vascular system at the time. For example, in an individual with increased vascular tone, the net effect of small amounts of epinephrine may be to reduce the mean blood pressure while increasing the heart rate and cardiac output. In an individual with a reduction in vascular tone, the mean blood pressure would be expected to increase, in addition to the reflex mechanisms, vascular output is integrated by the central nervous system,

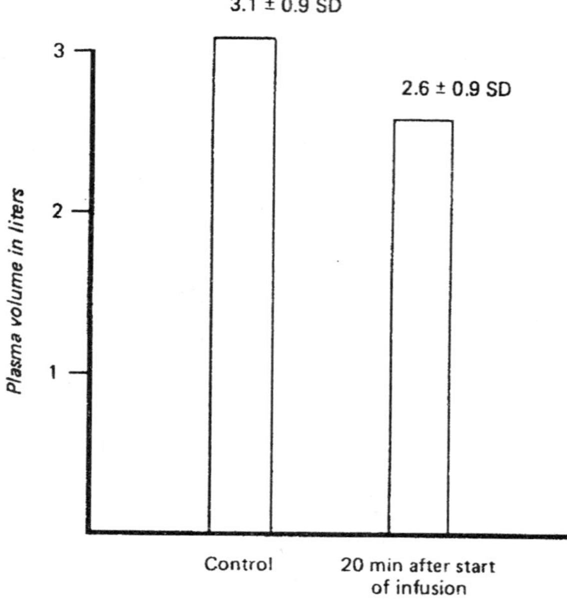

Fig. 6.6. Changes in plasma volume produced by infusion of norepinephrine for 20 minutes in a dose sufficient to increase mean arterial pressure from 96±10 mm Hg to 150±16 mm Hg.

so that, under appropriate circumstances, one vascular bed may be dilated while others remain unchanged. The central organization of the sympathetic nervous system is such that its ordinary regulatory effects are quite discrete—in contrast to periods of stress, when stimulation may be rather generalized and accompanied by release of catecholamines into the circulation. The infusion of catecholamines leads to a rapid reduction in plasma volume, presumably to accommodate to the reduced volume of the arterial and venous beds.

Extravascular smooth muscle

The catecholamines also regulate the activity of smooth muscle in tissues other than blood vessels. These effects include relaxation and contraction of uterine myometrium, relaxation of intestinal and bladder smooth muscle, contraction of the smooth muscle in the bladder and intestinal sphincters, and relaxation of tracheal smooth muscle governing bronchodilatation and pupillary dilatation.

Metabolic effects of catecholamines

The catecholamines increase oxygen consumption and heat production. Although the effects appear to be mediated by the β receptor, the mechanism is unknown. The catecholamines also regulate glucose and fat mobilization from storage depots. Glycogenolysis in heart muscle and in liver leads to an increase in available carbohydrate for utilization. Stimulation of adipose tissue leads to lipolysis and the release of free fatty acids and glycerol into the circulation for utilization at other sites. In humans, these effects seem to be mediated by the β receptor. The plasma levels of catecholamines required to produce some cardiovascular and metabolic effects in humans.

The catecholamines have effects on water, sodium, potassium, calcium, and phosphate excretion in the kidney. However, the mechanisms and the significance of these changes are not clear.

Role of catecholamines in the regulation of hormone secretion

The sympathetic nervous system plays an important role in the regulation and integration of hormone secretion at 2 levels. Centrally, norepinephrine and dopamine play important roles in the regulation of secretion of the anterior pituitary hormones. Dopamine, for example, has been identified as the prolactin-inhibitory factor, and the hypothalamic-releasing hormones appear to be under sympathetic nervous system control. Peripherally, the secretion of renin by the juxtaglomerular cells of the kidney is regulated by the sympathetic nervous system via the renal nerves and circulating catecholamines. The catecholamines release renin via a β receptor mechanism. The B cell of the pancreatic islets is stimulated by activation of the β receptors in the presence of α-adrenergic blockade. However, the dominant effect of norepinephrine or epinephrine is inhibition of insulin secretion mediated by the α receptor. Similar effects have been observed in the secretion of glucagon by pancreatic A cells. The catecholamines have also been found to increase the release of thyroxine, calcitonin, parathyroid hormone, and gastrin by a β receptor-mediated mechanism.

Other Hormones

In addition to the catecholamines, the chromaffin cells of the adrenal medulla and peripheral sympathetic neurons synthesize and secrete opiatelike peptides, including met- and leu-enkephalin. They are stored in the large, dense-cored vesicles with the catecholamines in the adrenal medulla and at sympathetic nerve endings. These peptides are also found in the terminals of the splanchnic fibers that innervate the adrenal medulla. The observation that naloxone increases plasma catecholamine levels suggests that these peptides may inhibit sympathetic activity.

Disorders of Adrenal Medullary Function

Hypofunction of the Adrenal Medulla

Hypofunction of the adrenal medulla alone probably occurs only in individuals receiving adrenocortical steroid replacement therapy following adrenalectomy. Such individuals with otherwise intact sympathetic nervous systems suffer no clinically significant disability. Patients with autonomic insufficiency, which includes deficiency of adrenal medullary epinephrine secretion, can be demonstrated to have minor defects in recovery from insulin-induced hypoglycemia. Patients with generalized autonomic insufficiency usually have orthostatic hypotension. When a normal individual stands, a series of physiologic

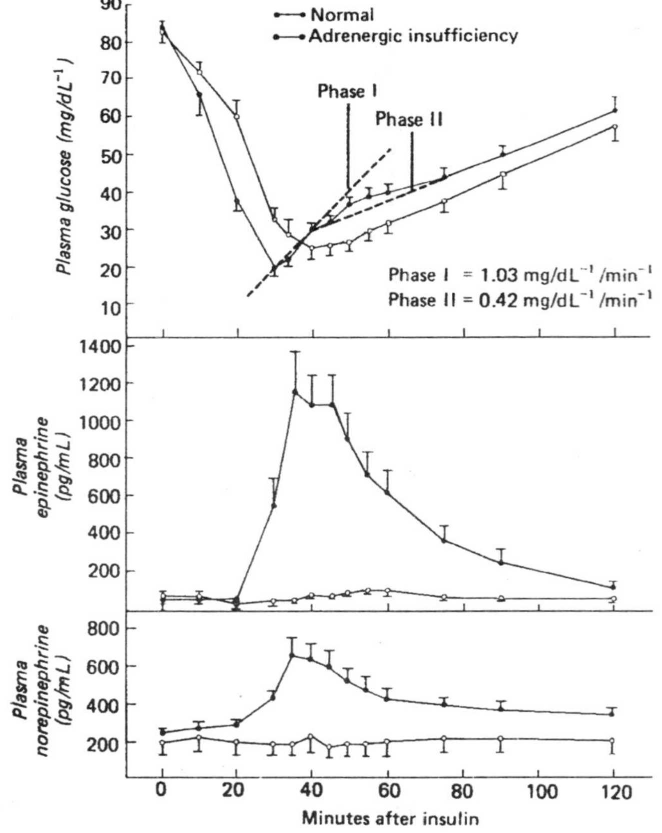

Fig. 6.7. Plasma glucose, epinephrine, and norepinephrine levels after insulin administration to 14 normal subjects and 7 patients with idiophathic hypotension with low low or absent epinephrine responses.

adjustments occur that maintain blood pressure and ensure adequate circulation to the brain. The initial lowering of the blood pressure stimulates the baroreceptors, which then activate central reflex mechanisms that cause arterial and venous constriction, increase cardiac output, and activate the release of renin and vasopressin. Interruption of afferent, central, or efferent components of this autonomic reflex results in autonomic insufficiency.

The treatment of symptomatic orthostatic hypotension is dependent upon maintenance of an adequate blood volume. If physical measures such as raising the head of the bed at night and using support garments do not alleviate the condition, pharmacologic measures may be used. Although agents producing constriction of the vascular bed, including ephedrine, phenylephrine, metaraminol, monoamine oxidase inhibitors, levodopa, propranolol, and indomethacin, have been used, volume expansion with fludrocortisone is the most effective treatment.

Hyperfunction of the Adrenal Medulla

The adrenal medulla is not known to play a significant role in essential hypertension. However, the role of the sympathetic nervous system in the regulation of blood flow and blood pressure has led to extensive investigations of its role in various types of hypertension. Some of the abnormalities observed, such as a resetting of baroreceptor activity, are thought to be secondary to the change in blood pressure. Others, such as the increased cardiac output found in early essential hypertension, have been thought by some investigators to play a primary role.

Catecholamines can increase blood pressure by increasing cardiac output, by increasing peripheral resistance through their vasoconstrictive action on the arteriole, and by increasing renin release from the kidney, leading to increased circulating levels of angiotensin II. Although many studies show an increase in circulating free catecholamine levels, evidence of increased sympathetic activity has not been a uniform finding.

Pheochromocytoma

Pheochromocytomas are tumors arising from chromaffin cells in the sympathetic nervous system. They release epinephrine or norepinephrine (or both)—and in some cases dopamine—into the circulation, causing hypertension and other signs and symptoms. It is estimated that 0.1% of patients with diastolic hypertension have pheochromocytomas. These tumors are found at all ages and in both sexes and are most commonly diagnosed in the fourth and fifth decades. Although uncommon, this disorder is important to diagnose, because undiagnosed it may be fatal in pregnant women during delivery or in patients undergoing surgery for other disorders. However, with early diagnosis and proper management, almost all patients with benign tumors recover completely.

Clinical Manifestations

Symptoms and signs

Although most patients with functioning tumors have symptoms most of the time, these vary in intensity and are perceived to be mainly episodic or paroxysmal by about half of the patients. Most patients with persistent hypertension also have superimposed paroxysms. A few patients are entirely free of symptoms and hypertension between attacks and give no evidence of excessive catecholamine release during these intervals. In rare instances, these tumors have been reported to occur without clinical manifestations.

Description of an attack

In patients with paroxysmal release of catecholamines, the symptoms resemble those produced by injections of epinephrine or norepinephrine, and the symptom complex is far more consistent than is suggested by the variability of patients' complaints. An episode usually begins with a sensation of something happening deep inside the chest, and a stimulus to deeper breathing is noted. The patient then becomes aware of a pounding or forceful heartbeat, caused by the β_1 receptor-mediated increase in cardiac output. This throbbing spreads to the rest of the trunk and head, causing headache or a pounding sensation in the head. The intense α receptor-mediated peripheral vasoconstriction causes cool, moist hands and feet and facial pallor. The combination of increased cardiac output and vasoconstriction causes marked elevation of the blood pressure when large amounts of catecholamines are released. The decreased heat loss and increased metabolism may cause a rise in temperature or flushing and lead to reflex sweating, which may be profuse and usually follows the cardiovascular effects that begin in the first few seconds after onset of an attack. The increased glycolysis and α receptor-mediated inhibition of insulin release cause an increase in blood sugar levels. Patients experience marked anxiety during all but the mildest attacks, and when episodes are prolonged or severe, nausea, vomiting, visual disturbances, chest or abdominal pain, paresthesias, and seizures can occur. A feeling of fatigue or exhaustion usually follows these episodes unless they are very mild or of short duration.

Frequency of attacks

In patients with paroxysmal symptoms, attacks usually occur several times a week or oftener and last for 15 minutes or less though they may occur at intervals of months or as often as 25 times daily and may last from minutes to days. As time passes, the attacks usually increase in frequency but do not change much in character. They are frequently precipitated by activities that compress the tumor, such as changes in position, exercise, lifting, defecation, or eating and by emotional distress or anxiety.

Variability of manifestations

Although the pattern of the manifestations described above can be elicited in almost all patients capable of clear communication, the variability of presenting complaints may be confusing and is sometimes misleading. Women whose episodes are first, noted around the time of menopause may be thought to be experiencing "hot flushes." The diagnosis may only be made when hormonal therapy has failed to alleviate the "hot flush" or the episode is observed and the blood pressure taken during the attack. When a pheochromocytoma causes hypertension late in pregnancy, it may be confused with preeclampsia. Other causes of increased sympathetic activity must be distinguished from pheochromocytomas.

Chronic symptoms

Patients with persistently secreting tumors and chronic symptoms usually experience the symptom complex described above periodically in response to transient increases in the release of catecholamines provoked by the same stimuli. In addition, in these patients the increased metabolic rate usually causes heat intolerance, increased sweating, and weight loss or (in children) lack of weight gain. The effects on glycogenolysis and insulin release can produce hyperglycemia and glucose intolerance, and patients may present with diabetes mellitus. Hypertension is usually present. Wide fluctuations of blood pressure are characteristic, and marked increases may be followed by hypotension and syncope. When pressure is elevated, postural hypotension is present. Typically, the hypertension does not respond to commonly used antihypertensive regimens, and such drugs as guanethidine and ganglionic blockers can induce paradoxical pressor responses. These patients, usually thin, have a forceful heartbeat, which is often visible and easily palpable. They feel warm, may have pallor of the face and chest, perspire, have cool and moist hands and feet, and prefer a cool room. A mass may be palpable in the abdomen or neck, and deep palpation of the abdomen may produce a typical paroxysm. Chronic constriction of the arterial and venous beds leads to a reduction in plasma volume in most of these patients. The inability to further constrict this bed upon arising causes the postural hypotension that is characteristically observed.

Familial Syndromes and Other Tumors

Pheochromocytoma may also occur as a heritable disorder, either alone or more commonly in association with other endocrine tumors. In one syndrome, multiple endocrine neoplasia (MEN) type II (Sipple's syndrome), the patient may also have a calcitonin-producing adenoma of the thyroid, a parathyroid hormone-producing adenoma of the parathyroid, or a pituitary adenoma. In the other group (MEN type III), pheochromocytomas occur in association with mucosal neuromas, which are numerous and small and are found around the mouth. Transmission of these disorders follows the pattern of an autosomal dominant gene with incomplete penetrance.

Pheochromocytomas occurring as part of the familial syndromes appear to be the expression of an as yet unidentified stimulus to tumor formation. In Sipple's syndrome, bilateral adrenal tumors, frequently producing epinephrine, are common and may develop from hyperplasia of the adrenal medulla. Medullary carcinoma in these patients is also commonly bilateral and associated with parafollicular or "C" cell hyperplasia. Multicentric tumors of the parathyroid gland and medullary hyperplasia are also quite typical and are found predominantly in patients with medullary carcinoma of the thyroid. The thyroid tumors have been reported to secrete ACTH, serotonin, prostaglandins, and kallikreins, which may contribute to the clinical manifestations in these patients.

Pathology

Location of tumor

Pheochromocytomas occur wherever chromaffin tissue is found. The adrenal medulla contains the largest collection of chromaffin cells. They are found in the organ of Zuckerkandl, which is very large in the fetus but is gradually replaced by fibrous tissue after delivery and is small in the adult. Chromaffin cells are also found in association with sympathetic ganglia, nerve plexuses, and nerves.

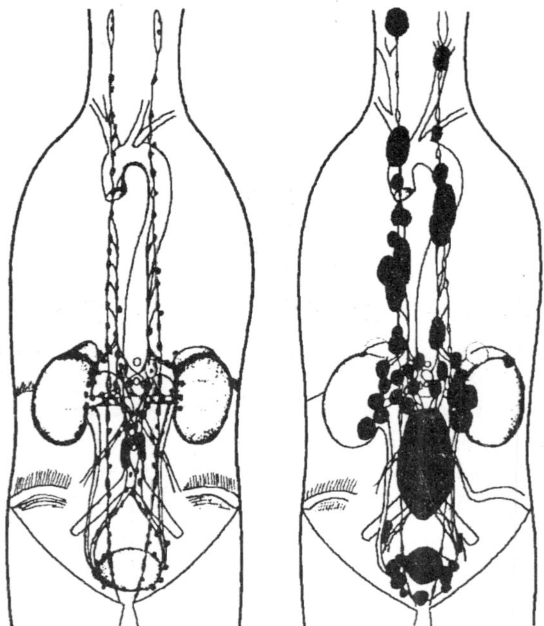

Fig. 6.8. Left: Anatomic distribution of extra-adrenal chromaffin tissue in the newborn. Right: Locations of extra-adrenal pheochromocytomas.

Over 95% of pheochromocytomas are found in the abdomen, and 85% of these are in the adrenal. Common extra-adrenal sites are near the kidney and in the organ of Zuckerkandl. Those found in the chest are in the posterior mediastinal area. The intracranial lesions reported are thought to be metastatic in origin. The tumors may be multicentric in origin, particularly when familial or part of the syndromes of multiple endocrine neoplasia and in children. Although fewer than 10% of adults have multiple tumors, they are found in about one-third of affected children.

Size of tumor tissue

Pheochromocytomas vary in size from less than 1 g to several kilograms. However, they are usually small, most weighing under 100 g. They are vascular tumors and commonly contain cystic or hemorrhagic areas. The cells tend to be large and contain typical catecholamine storage granules similar to those in the adrenal medulla. Multinucleated cells, pleomorphic nuclei, mitoses, and extension into the capsule and vessels are sometimes seen but do not indicate that the tumor is malignant. Fewer than 10% of pheochromocytomas are malignant, and these can be recognized either during surgery, by the finding of extensive local infiltration or metastases, or later, because of postsurgical local recurrence or metastases.

Adrenal medullary hyperplasia

This has also been described as the cause of an indistinguishable clinical picture and is found in the syndrome of multiple endocrine neoplasia.

Complications

Patients with persistent symptoms and hypertension may develop hypertensive retinopathy or nephropathy. Injection of catecholamines leads to a diffuse myocarditis, and postmortem studies indicate that there are a significant number of patients with pheochromocytomas who also have lesions characterized by focal degeneration and necrosis of myocardial fibers with infiltration of histiocytes, plasma cells, and other signs of inflammation. Platelet aggregates and Fibrin deposition are found in pulmonary arterioles. These changes may be associated with abnormal findings on the ECG but may not become apparent until the patient is exposed to marked cardiovascular stress and heart failure results. In patients harboring pheochromocytomas for long periods, the serious sequelae of hypertension are observed. Cerebrovascular accidents, congestive heart failure, and myocardial infarction have been observed.

Differential Diagnosis

The diagnosis of pheochromocytoma should be considered in all patients with paroxysmal symptoms; in children with hypertension; in adults with severe hypertension not responding to therapy; in hypertensive patients with diabetes or hypermetabolism: in patients with hypertension in whom symptoms resemble those described above or can be evoked by exercise, position change, emotional distress, or antihypertensive drugs such as guanethidine and ganglionic blockers; and in patients who become severely hypertensive or go into shock during anesthesia, surgery, or obstetric delivery. Patients who have disorders sometimes associated with pheochromocytomas (neurofibromatosis, mucosal adenomas, von Hippel's disease, medullary carcinoma of the thyroid) and those with first-degree relatives who have pheochromocytoma should be investigated.

In these and other disorders in which elevated plasma catecholamine levels are under sympathetic nervous system control, it may be possible to reduce catecholamine release by autonomic blockade or sympathetic suppression. Pentolinium and clonidine have been used for this purpose. The reliability and safety of such procedures have not been established.

Ganglioneuromas, which are usually small, well-differentiated tumors arising from ganglion cells, and neuroblastomas, which are highly malignant tumors arising from more primitive sympathoblastic cells, can produce catecholamines and present a similar clinical picture. Dopamine is usually the major active catecholamine produced and leads to elevation of urinary homovanillic acid.

Diagnostic Tests and Procedures

Assay of catecholamines and their metabolites has markedly simplified the diagnosis of this disorder. Operative exploration for pheochromocytoma should not be done in the absence of chemical confirmation of the diagnosis. Using currently available methods, it is possible to avoid unnecessary surgery and to successfully locate a tumor in almost all patients.

Hormone assay

In patients with continuous hypertension or symptoms, levels of plasma or urine catecholamines and their metabolites are usually clearly increased. Therefore, in the selection of a particular assay, it is more important that the test be performed well by the laboratory than that a particular substance be measured. A reliable assay of the catecholamines, metanephrines, or vanillylmandelic acid (VMA) is usually sufficient to confirm the diagnosis. Patients with large tumors may excrete disproportionately greater amounts of catecholamine metabolites, because the amines can be metabolized by enzymes in the tumor cells prior to their release. Malignant tumors may release large amounts of dopamine, leading to the excretion of large amounts of homovanillic acid in the urine. It is important that the appropriate assay procedure be chosen to avoid misleading results and that drugs and foods that interfere with these assays be eliminated.

In patients having brief and infrequent paroxysms with symptom-free intervals, confirmation of the diagnosis may be more difficult. Although large amounts of catecholamines are produced during the brief episode, the total amount excreted during the 24-hour urine collection period may not be clearly abnormal—in contrast to patients whose tumors secrete continuously. The latter group will accumulate larger amounts of catecholamines and metabolites even though secretion rates are lower and symptoms are less severe because of their continuous secretion. Therefore, sampling of blood or timed urine collections during a carefully observed episode may be necessary to confirm the diagnosis.

Glucagon test

In patients with infrequent episodes, it may be useful to induce a paroxysm. This is not often necessary and should not be done in patients who have angina, visual changes, or other severe symptoms during spontaneous attacks, and phentolamine should be available to terminate the induced episode. Injection of 1 mg of glucagon intravenously will induce an attack in more than 90% of patients with pheochromocytoma. In those instances in which glucagon has failed to evoke a paroxysm and the clinical suspicion is very strong, histamine given intravenously in doses of 25-50 μg may also be tried. However, histamine injection is associated with flushing and a brief but severe episode of headache. In general, the provoked episodes are no more severe than those occurring spontaneously.

In the occasional patient in whom the chemical tests are inconclusive, it may be useful or convenient to institute therapy with phenoxybenzamine over a period of 1-2 months to observe effects both

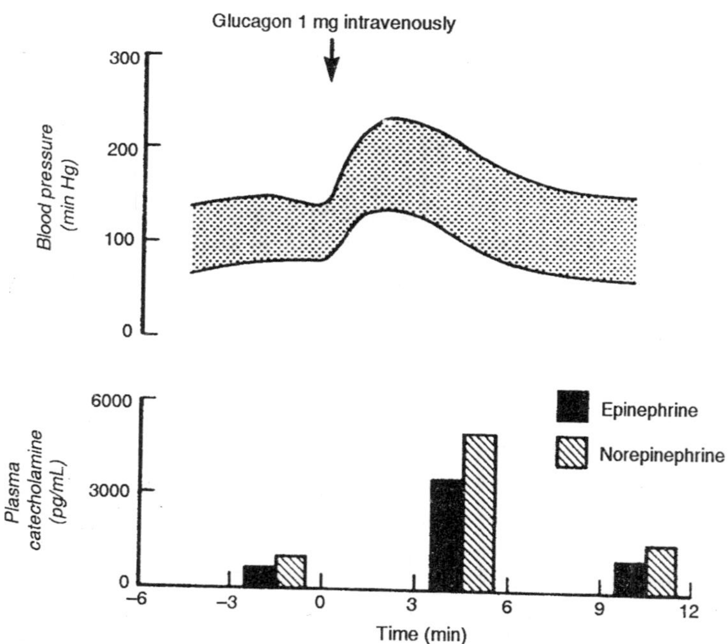

Fig. 6.9. Blood pressure and plasma catecholamine levels following glucagon injection in a patient with a paroxysmally secreting phenochromocytoma.

on the nature and frequency of attacks and on the blood pressure. A salutary effect will sometimes be observed for a few weeks but seldom longer than that in the absence of a pheochromocytoma. A good response indicates the need for reappraisal of the patient.

Screening Patients for Pheochromocytoma

In addition to screening patients with typical clinical manifestations, assays for catecholamines and their metabolites can be used to screen other patients at high risk. When the presence of an undetected pheochromocytoma is particularly dangerous, as in pregnancy or prior to surgery, screening should be considered even when clinical manifestations are less typical.

Localization of Tumors

When the diagnosis has been established, the tumor must be located in order to facilitate its surgical removal. Although the larger tumors can usually be easily located by conventional radiographic procedures (urography and arteriography) and sonography, the smallest tumors, particularly when located outside the adrenal, may require special techniques. Early experience with CT scanning has given excellent results. Only the smallest tumors or those shielded by clips and other

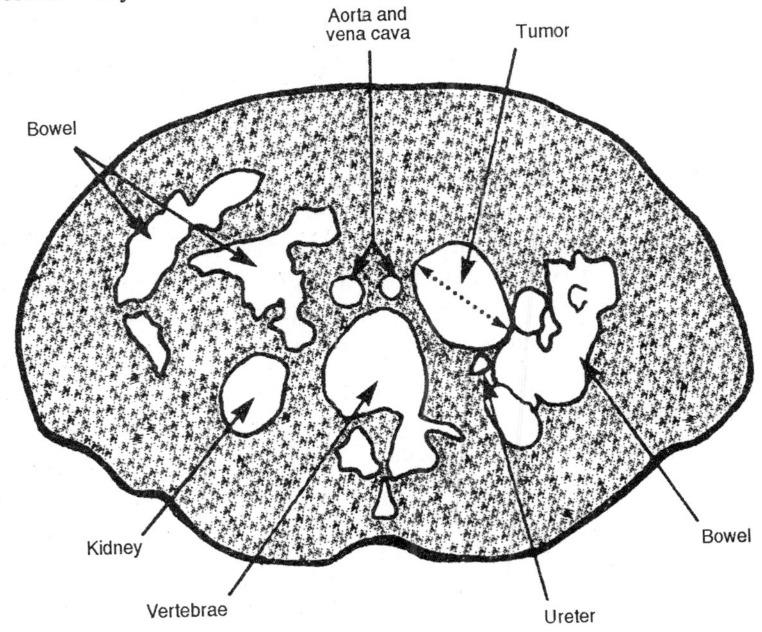

Fig. 6.10. The diagram identifies many of the visible structures.

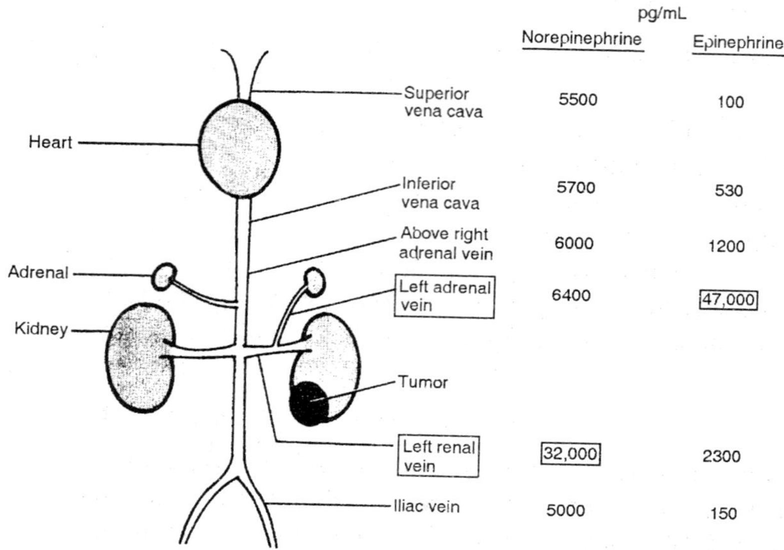

Fig. 6.11. Plasma norepinephrine and epinephrine levels in samples of blood obtained from the vena cava.

metal objects from previous surgery have been elusive. Analysis of blood samples obtained via percutaneous venous catheterization can be of great value in locating small tumors in unusual locations. Because of a small risk of complications and the discomfort and cost of this procedure, it should be used only after simpler methods fail.

A scintigraphic procedure for the localization of pheochromocytomas has been recently described. [^{131}I]Metaiodobenzylguanidine (MIBG) is injected into the patient, resulting in detectable images 24—72 hours later. Early experience indicates that very small tumors can be safely and conveniently located by this method. Although not all pheochromocytomas will produce detectable images, the specificity of the method seems high, and it has been possible to locate tumors not shown by CT scanning.

Management

Optimal management of patients with pheochromocytomas requires an understanding of the pathophysiology produced by excessive catecholamines and an acquaintance with the action of adrenergic antagonists and other drugs used in the treatment of these patients.

Adrenergic antagonists

As soon as the diagnosis has been confirmed, therapy with adrenergic antagonists should be instituted. Treatment is directed toward reduction of symptoms, lowering of blood pressure, and amelioration of paroxysms occurring spontaneously or induced by studies undertaken to localize the tumors. Such treatment will allow expansion of the vascular bed and plasma volume and reduce the amount of transfused blood required for maintenance of blood pressure during surgery.

Although only a few days of therapy are required for preoperative study and preparation of most patients for surgery, prolonged medical therapy is advantageous in patients who have had recent myocardial infarctions, those with electrocardiographic or clinical evidence of catecholamine cardiomyopathy, or those in the last trimester of pregnancy. They can be maintained on this regimen until the baby is delivered or the complication resolved, at which time the tumor can be removed.

Drugs used

Agents commonly used in therapy include phentolamine and phenoxybenzamine. Propranolol is occasionally helpful.

Phenoxybenzamine

In patients with persistent hypertension or frequent paroxysms, phenoxybenzamine (Dibenzyline), a noncompetitive α-adrenergic antagonist with a prolonged effect, is indicated. Treatment is begun with doses of 20-40 mg/d orally and can be increased by 10—20 mg every 1-2 days until the desired effect is achieved. Postural hypotension may be marked at the beginning of therapy. Completely normal blood pressures (< 140/90 mm Hg) may not be achieved and are not required. A dose of 60-80 mg daily is usually adequate, but 2-3 times that amount may be necessary. When marked tachycardia or arrhythmias occur prior to or during surgery, small doses of propranolol may be required. Patients with infrequent paroxysms and an absence of interval manifestations can be treated in a similar manner.

Prazosin

Rapid titration of the dose may be difficult using phenoxybenzamine, because of its long half-life (36 hours). However, prazosin (Minipress), a recently introduced a antagonist with a shorter duration of activity, could prove to be useful in such patients.

Preparation for surgery

Preparation of the patient in this manner minimizes the hazards of anesthesia and surgery. The patient's blood pressure and ECG should be continuously monitored, and phentolamine, propranolol, and sodium nitroprusside can be used to reduce blood pressure if necessary.

Suspected intra-abdominal tumors are usually approached through a transabdominal incision, since this allows exploration of the adrenals, sympathetic ganglia, bladder, and other pelvic structures. When bilateral adrenal tumors are found and the adrenals removed, adrenocortical steroid replacement is required.

Postoperative care

When the tumor is removed, the blood pressure usually falls to about 90/60 mm Hg. Persistence of low blood pressures or poor peripheral perfusion may require blood volume expansion with whole blood, plasma, or other fluids as indicated. Pressor therapy is not usually required and should not be substituted for volume expansion.

Lack of a fall in pressure at the time of tumor removal (even in the presence of adrenergic blockade) indicates the presence of additional tumor tissue.

In patients with tumors producing persistent hypertension, the initial fall in blood pressure may be followed by an elevation of blood pressure in the postoperative period. However, the accompanying symptoms of sympathetic stimulation are gone, and the pressure returns to normal over the next few weeks. If the blood pressure remains elevated but the patient is otherwise asymptomatic, another cause for the elevated blood pressure should be considered. Essential hypertension and renal vascular hypertension have been reported in several patients harboring pheochromocytomas.

Treatment of malignant tumors

Patients with nonresectable malignant tumors or metastases or those who for other reasons are not amenable to successful surgical treatment can be managed medically for prolonged periods. Phenoxybenzamine and perhaps prazosin can be used chronically as described above. Patients with malignant tumors have also benefited symptomatically from treatment with α-methylmetatyrosine, an inhibitor of tyrosine hydroxylase, the rate-limiting enzyme in the biosynthetic process. Although some patients with malignant pheochromocytomas die early because of disseminated disease, there are long-term survivors. The most common site of metastases is the skeleton, and bone lesions tend to respond well to radiation therapy. The use of chemotherapy, alone or in combination with radiation, for soft tissue lesions has been disappointing.

7

THYROID GLAND

Thyroid hormones (thyroxine, T_4; triiodothyronine, T_3) play a vital role in fetal development, and throughout life they influence metabolic processes in almost all tissues. The thyroid gland is the only significant source of T_4. While the gland does produce some T_3, which is biologically more active than T_4, in humans, most T_3 is produced extrathyroidally from T_4.

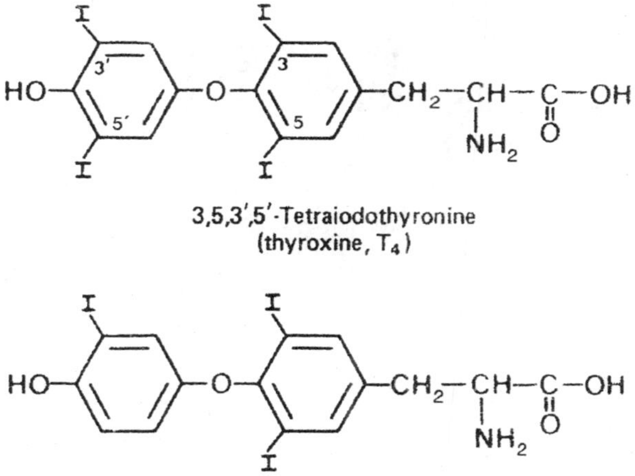

3,5,3′,5′-Tetraiodothyronine (thyroxine, T_4)

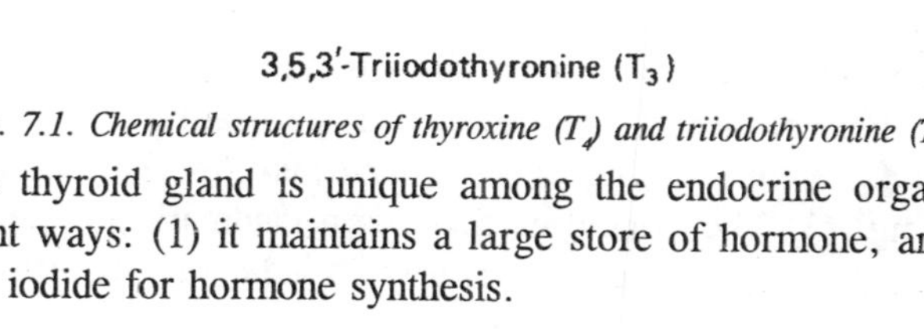

3,5,3′-Triiodothyronine (T_3)

Fig. 7.1. Chemical structures of thyroxine (T_4) and triiodothyronine (T_3).

The thyroid gland is unique among the endocrine organs in 2 important ways: (1) it maintains a large store of hormone, and (2) it requires iodide for hormone synthesis.

Anatomy of the Thyroid

Gross Anatomy

The thyroid consists of a left and right lobe connected (in humans) by an isthmus at the approximate level of the cricoid cartilage. The adult thyroid weighs 15-20 g, and assessment of its size is important in clinical evaluation of thyroid function. The thyroid is derived from the embryonic thyroglossal duct, which is a midline invagination at the junction of the anterior two-thirds and posterior one-third of the tongue. A remnant of the thyroglossal duct may be observed as a pyramidal thyroid lobe, inserting into the isthmus or into the medial aspects of one of the lobes.

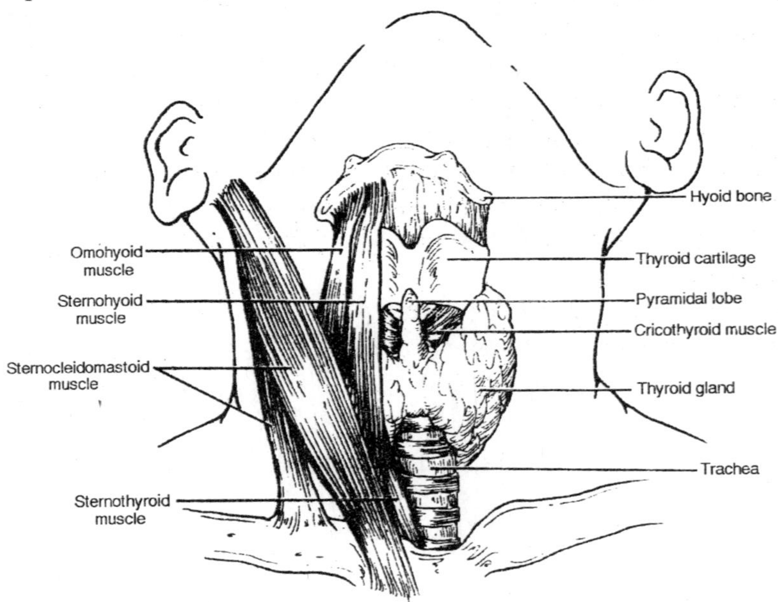

Fig. 7.2. Gross anatomy of the human thyroid (anterior view).

In palpating the thyroid gland, it is important to note the following: (1) The thyroid lobes are largely covered by the sternohyoid and sternothyroid muscles. Enlargement of the thyroid is limited superiorly by the sternothyroid attachment to the thyroid cartilage, but the thyroid gland, when enlarged, can extend superolaterally beneath the sternocleidomastoid. (2) The thyroid moves upward with the larynx on swallowing. (3) The cricoid cartilage is a landmark for the isthmus. (4) Pyramidal lobe enlargement is a good indication of diffuse rather than local abnormality of the thyroid.

Histology

The basic functional unit of the thyroid is the follicle, a hollow sphere of cells, 15-500 μm in diameter, surrounded by a basement membrane. The wall of the follicle is a single layer of thyroid cells. A few adjacent parafollicular cells (C cells) are larger, clearer on staining, and secrete calcitonin. The follicular cells are cuboidal when quiescent and columnar when active. The follicular lumen contains colloid, a viscous gel that is primarily a store of thyroglobulin secreted by the thyroid cells. This store is sufficient for about 100 days of normal thyroid hormone secretion. The follicles are surrounded by a rich capillary network. Blood flow through the thyroid is very high (about 5 mL/g/min) and may be audible with a stethoscope (bruit) when the thyroid is overactive and flow is increased. A sympathetic and parasympathetic nerve network also surrounds the follicles. Their significance is unclear, however, and they do not appear to play a major role in the regulation of thyroid hormone synthesis and secretion.

Iodide Metabolism

Adequate ingestion of iodide is a prerequisite for the normal synthesis of thyroid hormones by the thyroid. While the dietary intake of iodide may vary widely, thyroidal secretion of iodine in the form of thyroid hormones is relatively stable. The homeostatic mechanisms at play will be discussed below. The major sources of dietary iodide are water, iodated bread, iodinated salt, and, increasingly, medications, disinfectants, and x-ray contrast material. In addition, kelp, available in health food stores, is being ingested with increasing frequency in the USA.

A typical daily dietary intake of iodide in the USA is now about 500 μg. Iodide is almost completely absorbed in the gastrointestinal tract, where it enters the inorganic iodide pool in the extracellular fluid. In the presence of normal renal function, inorganic iodide in the extracellular fluid is rapidly cleared, with a half-life of about 2 hours. Besides dietary iodide, there are 2 other smaller sources contributing to the extracellular fluid iodide pool: (1) iodide released by the deiodination of thyroid hormones in the peripheral tissues (about 60 μg/d) and (2) the release ("leak") of inorganic iodide by the thyroid (about 10-50 μg/d).

The only 2 significant pathways of iodide clearance from the extracellular fluid are the kidneys and the thyroid. Since the thyroid is able to regulate the net amount of iodide it clears from the extracellular fluid, taking up only as much as is needed (about 75 μg/

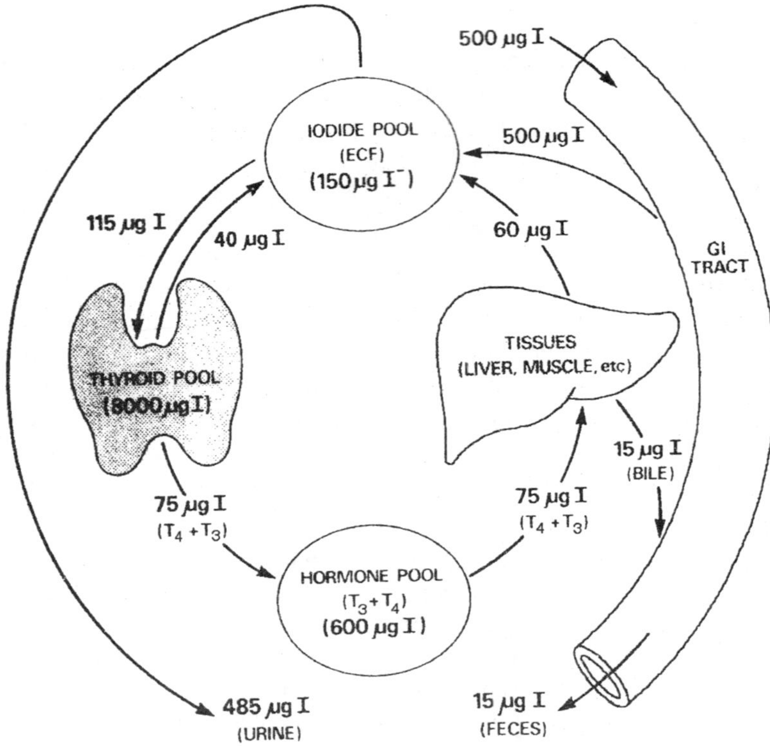

Fig. 7.3. Iodine metabolism.

d) for thyroid hormone synthesis, it follows that when the dietary intake of iodide increases, the fractional uptake of extracellular fluid iodide by the thyroid decreases and the proportionate urinary excretion of iodide increases. Conversely, when iodide intake is reduced, the proportionate uptake of extracellular fluid iodide by the thyroid is increased. This is clinically important, because the percentage uptake of radioactive iodide by the thyroid is a useful index of thyroid function. In the USA, particularly since the iodation of bread, the radioactive iodide uptake (RAIU) by the thyroid gland in normal subjects has decreased from about 20-50% of the ingested dose at 24 hours to about 10-25%. RAIU is generally higher in hyperthyroidism and lower in hypothyroidism, although there are exceptions. Measurement of urinary iodide excretion may be useful clinically if it is suspected that the patient has ingested excess iodide.

The largest iodine pool in the body is the slowly turning over thyroid pool (approximately 1% per day: faster in hyperthyroidism).

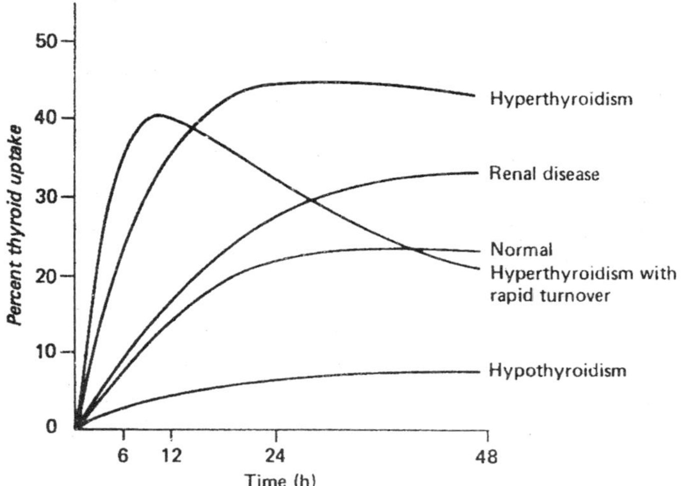

Fig. 7.4. Typical radioactive iodine uptake (RAIU) curves over 48 hours in patients with normal and abnormal thyroid function.

This pool is almost entirely organic iodine in the form of iodinated thyroglobulin stored as colloid within the thyroid follicular lumen.

The thyroid secretes about 75 μg of organic iodine per day in the form of thyroid hormones (chiefly T_4, with a small amount of T_3). Secreted thyroid hormones enter a pool of thyroid hormones, mainly intravascular, bound to thyroid hormone-binding proteins. The iodine content of this hormonal pool is about 600 μg. Cellular uptake of thyroid hormones from this pool is approximately 75 μg of iodine per day in the euthyroid individual. Of this 75 μg of thyroid hormone iodine, about 60 μg reenters the extracellular fluid iodide pool following intracellular enzymatic deiodination of the thyroid hormones. The remainder (15 μg) is conjugated in the liver to the glucuronide and sulfate forms and excreted in the bile and subsequently in the stool. In humans, there is very little enterohepatic recirculation of thyroid hormone.

Thyroid Hormone Synthesis

Thyroglobulin

Thyroglobulin, the precursor of all thyroid hormones, is a very large glycoprotein molecule that is the major protein in the follicular luminal colloid. About 75% of the weight of the entire thyroid gland consists of this protein, though the proportion varies widely depending on the physiologic demands on the gland.

Thyroglobulin is a heterogeneous substance that exists in multiple forms. The most prevalent form and the major precursor for thyroid hormones is a 19S molecule with a molecular weight of 660,000. Ultracentrifugation of thyroglobulin also reveals a small fraction with a sedimentation coefficient of 27S, believed to be a dimer of 19S thyroglobulin. The 19S molecule can in turn dissociate into two 12S components, particularly when the thyroglobulin is poorly iodinated. There is controversy over whether thyroglobulin is synthesized as a whole 19S molecule or whether subunits are first synthesized and then subsequently combined. The most prevalent theory, based on the translation of thyroglobulin messenger RNA (mRNA), suggests that thyroglobulin is synthesized as the 12S subunit. However, recent evidence suggests that thyroglobulin mRNA may code directly for the entire 19S molecule.

Human thyroglobulin contains approximately 110 tyrosine residues. It is these residues that are iodinated and ultimately form the thyroid hormones. The relative number of tyrosines in thyroglobulin is not higher than in many other proteins. The propensity of thyroglobulin for iodination is determined instead by its proximity to a very efficient iodination mechanism in the thyroid cell as well as by its tertiary structure, which facilitates iodination. Only a small proportion of the tyrosine residues in thyroglobulin—those that are closest to the surface of the globular molecule—are available for iodination. In geographic areas of iodine sufficiency, a thyroglobulin molecule will contain an average of about 7 molecules of monoiodotyrosine (MIT) and 5 of diiodotyrosine (DIT). In addition, an average of 2 molecules of T_4 will be present, each representing the coupling of 2 molecules of DIT. On average, every third molecule of thyroglobulin will contain one molecule of T_3, formed by the coupling of one DIT with one MIT.

The degree of thyroglobulin iodination varies depending upon the availability of iodide and the efficiency of the iodinating mechanism. Poorly iodinated thyroglobulin contains a relatively high MIT/DIT ratio, which in turn favours increased T_3 synthesis relative to T_4. This is presumably of adaptive significance in that during periods of iodide insufficiency, available iodide is utilized more efficiently by synthesizing a more active hormone. In addition, there is evidence that decreased thyroglobulin iodine content produces a more loosely folded molecule that is more susceptible to proteolysis within the thyroid and will therefore also lead to greater efficiency in the generation of biologically active thyroid hormones.

The mechanism of synthesis of thyroglobulin is the same as that of other glycoproteins. After transcription and processing of thyroglobulin mRNA has taken place, the mRNA is translated by ribosomes in the rough endoplasmic reticulum. The thyroglobulin polypeptide chain is extruded into the lumen of the endoplasmic reticulum. During transport to the Golgi apparatus, thyroglobulin is progressively glycosylated. The molecules are packaged into exocytotic vesicles in the Golgi apparatus. These vesicles fuse with the plasma membrane at the apical border of the follicular cell and release their contents into the follicular lumen. The membrane incorporated into the plasma membrane by this process is believed to recycle by the formation of endocytotic vesicles during the process of hormone secretion.

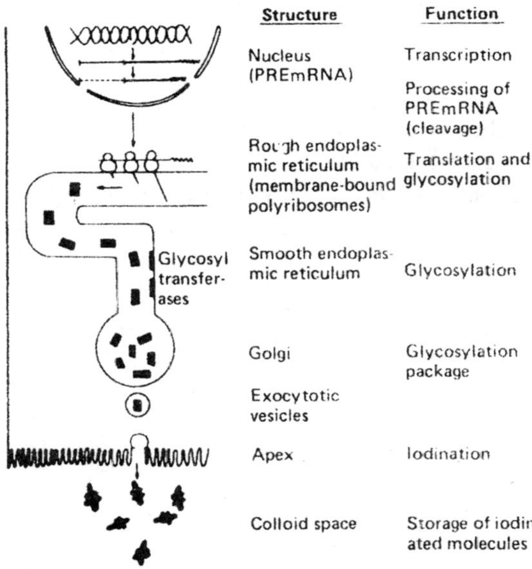

Fig. 7.5. Synthesis of thyroglobulin.

The exact site of thyroglobulin iodination is not established with certainty, but evidence suggests that iodination occurs at the apical (follicular) surface of the thyroid cell. This is based on the autoradiographic localization of radioiodinated thyroglobulin a few seconds after exposure of the thyroid to radioactive iodine. In addition, the enzyme necessary for thyroglobulin iodination can be demonstrated at the site by histochemical means.

The process of thyroglobulin biosynthesis and exocytosis into the follicular lumen is believed to be under the dominant control of TSH.

Evidence suggests that TSH stimulates both the transcriptional and translational processes, resulting in thyroglobulin synthesis. TSH also rapidly stimulates the exocytosis (extrusion into the follicular lumen) of preformed thyroglobulin.

Thyroglobulin is not the only protein that may be iodinated within the thyroid gland, and other proteins such as albumin may occasionally be iodinated. This, however, occurs to a significant degree only in the abnormal thyroid gland. Because other molecules lack the unique tertiary structure of thyroglobulin necessary for the efficient coupling of iodotyrosines, their iodination does not lead to adequate biologically active thyroid hormone synthesis. Excess iodoalbumin production may be detected by a disproportionately high organic iodide concentration in serum relative to the amount of thyroid hormone present.

Iodide Transport

Iodide is actively transported ("trapped") from the extracellular fluid into the thyroid follicular cell. Once within the cell, inorganic iodide may either be incorporated into protein ("organified") or diffuse back into the extracellular fluid. At equilibrium, the inorganic iodide concentration ratio between the thyroid cell and the serum (T:S iodide gradient) represents the balance between iodide influx and efflux. It is important to realize that organic iodine, such as iodinated thyroglobulin, does not contribute to this iodide gradient. In the normal thyroid gland, inorganic iodide is organified almost immediately upon entering the cell, so that a high T:S inorganic iodide gradient is not detectable.

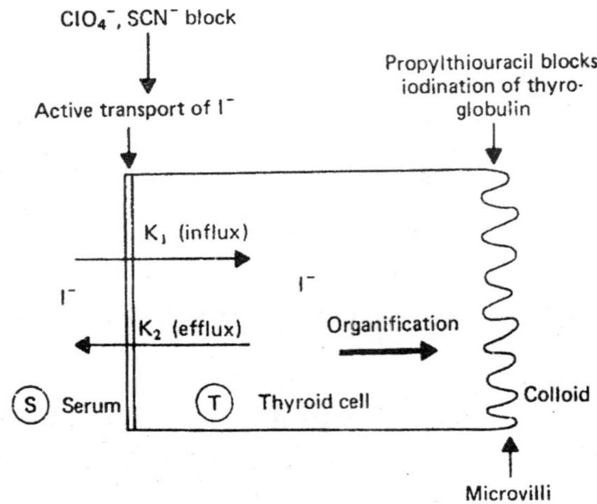

Fig. 7.6. The iodide transport mechanism in the thyroid cell.

However, if iodide organification is blocked, a T:S iodide ratio as high as 500:1 has been observed in the mouse thyroid *in vitro*.

Other structurally related anions such as SCN^-, BF_4^-, NO_3^-, and ClO_4^- competitively inhibit iodide transport. The latter (perchlorate) is used clinically on occasion to evaluate the iodide organification mechanism. If the thyroidal iodide organification mechanism is inefficient, inorganic iodide transported into the cell will accumulate. Perchlorate given a few hours after administration of a tracer amount of radioiodide blocks further iodide influx, but passive diffusion of inorganic iodide from the cell back into the extracellular fluid continues. Patients with faulty organification mechanisms have an abnormal "discharge" of radioactivity from the thyroid as it is released back into the extracellular fluid. This is detected by a decrease in radioactivity in the neck. The normal thyroid discharges little of the administered radioiodine with ClO_4^-, because it is readily organified and no longer able to leak out into the serum.

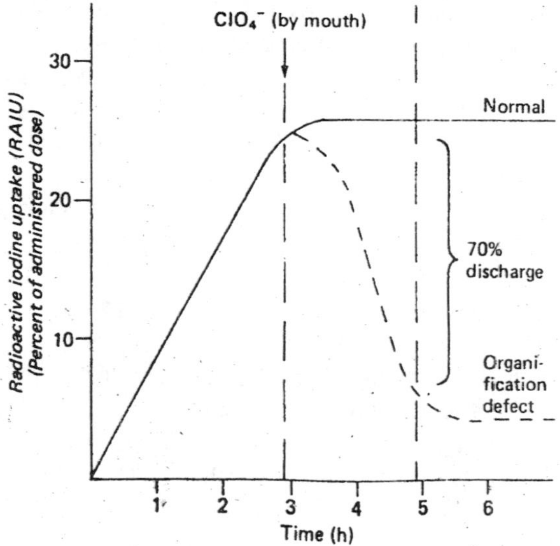

Fig. 7.7. Perchlorate discharge of thyroidal inorganic iodine.

Tissues other than the thyroid also have an active transport mechanism for iodide. These include gastric mucosa, salivary glands, and the choroid plexus. However, unlike the thyroid, these tissues have a minimal capacity for iodide organification and storage of organic iodine. The active process of I^- concentration against an electrochemical gradient is energy-dependent, requiring oxidative metabolism. Because cardiac glycosides inhibit thyroidal I^- transport, there is much interest in the possible role of Na^+-K^+ ATPase activity in the iodide transport mechanism. Sodium increases I^- influx and inhibits I^- efflux. However, the exact relationship between this enzyme, sodium transport, and the iodide transport mechanism is presently unclear.

Iodide Organification

Once within the thyroid cell, inorganic iodide is rapidly oxidized by thyroid peroxidase in the presence of H_2O_2 into a reactive intermediate that is then incorporated into the tyrosine residues of acceptor proteins, mainly thyroglobulin. Thyroid peroxidase is a membrane-bound hemoprotein that is relatively unstable and therefore difficult to purify. The exact nature of the reactive oxidized iodide intermediate that binds to acceptor proteins is unknown. Although I_2 (iodine) has been considered, more acceptable alternatives include a sulfenyl iodide or possibly the I° free radical.

The exact mechanism of thyroid H_2O_2 generation is also unknown, but it is probably dependent on reduced pyridine nucleotide oxidation. NADPH-cytochrome C reductase is believed to be involved in thyroidal H_2O_2 production in that a specific antibody against this enzyme is able to inhibit thyroglobulin iodination in vitro by a thyroid paniculate (microsomal) preparation.

As mentioned previously, the site of thyroglobulin iodination is most likely to be apical plasma membrane. This is based on electron-microscopic autoradiography of sections of thyroid performed within seconds after the administration of radioactive iodide. That this process occurs at the interface between the thyroid cell and the colloid is consistent with the fact that thyroid peroxidase is membrane-bound and with the direct localization of enzyme activity in apical membrane. Still being investigated, however, are the role of the thyroid peroxidase-like material that is demonstrable histochemically within the interior of the thyroid cell (such as in the Golgi apparatus) and the possible role in iodide organification of multiple pools of intracellular iodide differing in their rates of turnover.

Iodotyrosine Coupling

Iodinated tyrosine residues in thyroglobulin combine to form iodothyronines. Thus, DIT couples with DIT to form thyroxine and MIT with DIT to form triiodothyronine. The exact mechanism by which this coupling reaction occurs is unknown. Two theories currently exist: intramolecular coupling and intermolecular coupling. In the former, 2 peptide-linked DIT molecules couple while still part of the thyroglobulin

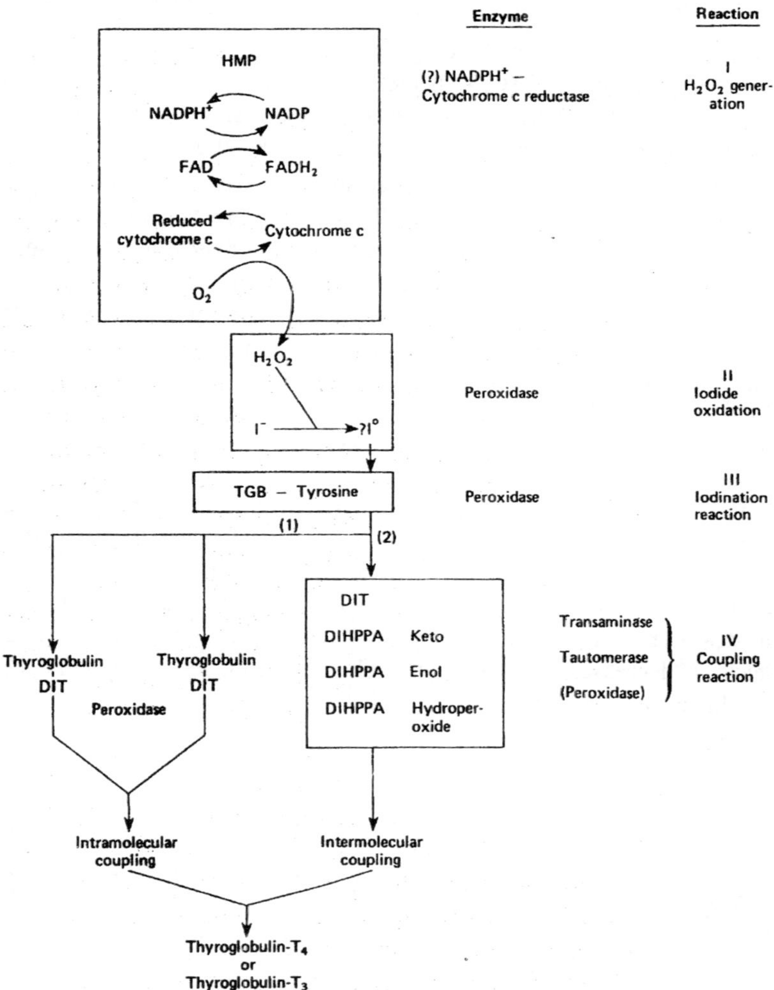

Fig. 7.8. Present concepts of thyroglobulin iodination.

polypeptide chain—a process that may involve generation of free radicals of DIT. In intermolecular coupling, there is evidence that DIT is liberated from thyroglobulin and in turn converted into its pyruvic acid analog diiodohydroxyphenylpyruvic acid (DIHPPA) by tyrosine transaminase. After conversion to its enol form and subsequent oxidation by H_2O_2 and thyroid peroxidase to the DIHPPA hydroperoxide, this intermediate then couples with DIT present within the thyroglobulin molecule to form thyroxine. While this issue remains unsettled, it is clear that thyroid peroxidase plays an important role in the coupling process as well as in initial iodotyrosine formation. Thyroid peroxidase therefore catalyzes all steps in the synthesis of iodothyronines. DIT may itself have a stimulatory effect on iodotyrosine coupling, but the mechanism by which this occurs is poorly understood.

A defect in the coupling mechanism results in diminished T_4 and T_3 formation from DIT and MIT. This may be related to inadequate concentrations of precursor iodotyrosines or to a defect in the enzymes responsible for coupling. As an example of the first possibility, in iodine deficiency, poorly iodinated thyroglobulin contains fewer iodotyrosine molecules overall and more MIT relative to DIT. There may therefore be insufficient DIT residues to permit adequate T_4 formation. As a corollary, there is a proportionate increase in T_3 production relative to T_4. This may represent an adaptive mechanism to environmental iodine deficiency in that T_3 is biologically more potent than T_4. While thyroid peroxidase is important in the coupling mechanism, it is not clearly established whether a deficiency in this enzyme or some abnormality in its action is responsible for the apparent coupling defect very rarely observed in the human thyroid. The difficulty in determining this is that decreased thyroid peroxidase activity also leads to decreased thyroglobulin iodination, with less MIT and DIT available for coupling.

A number of related drugs are used clinically to treat hyperthyroidism by decreasing thyroid hormone synthesis. The thionamide drugs, including methimazole and propylthiouracil, inhibit thyroid peroxidase-catalyzed iodination of thyroglobulin. These drugs may also specifically inhibit the iodotyrosine coupling reaction independently of inhibition of iodotyrosine generation. Propylthiouracil, unlike methimazole, inhibits iodotyrosine coupling in vitro.

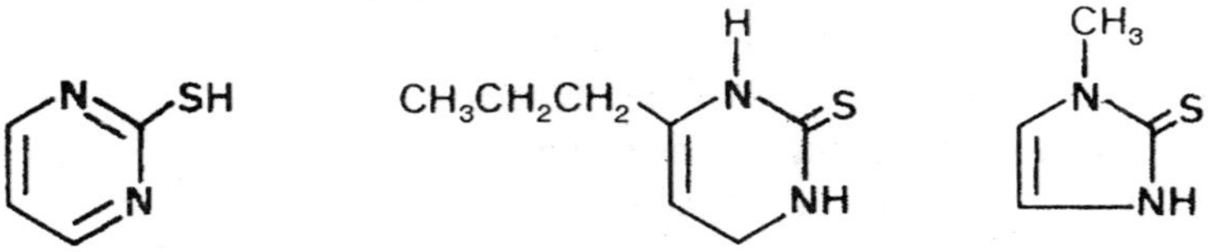

Fig. 7.9. Thiocarbamide inhibitiors of thyroidal iodide organification.

Thyroid Hormone Secretion

Thyroglobulin is stored extracellularly in the follicular lumen. Therefore, as a prerequisite for thyroid hormone secretion into the blood, thyroglobulin must first reenter the thyroid cell and undergo proteolysis. A small quantity of intact thyroglobulin enters the circulation, some by way of the lymphatics.

Autoradiographic studies show that newly iodinated thyroglobulin is stored adjacent to the epithelial border. Continuing synthesis may produce concentric lamellae of colloid with older thyroglobulin in the center. It follows, therefore, that the most recently iodinated thyroglobulin may be the first to be secreted. This has been called the "last come, first served" principle. Not all follicles in the thyroid are of the same size, and their thyroglobulin turnover rates vary considerably. The smaller follicles turn over their contents at a faster rate. Processing of thyroglobulin by the thyroid cell and secretion of the thyroid hormones into the circulation occur in the following sequence: (1) Pseudopods from the apical cell surface extend into the colloid in the follicular lumen, and large colloid droplets enter the cytoplasm by endocytosis. Each colloid droplet is enclosed in membrane derived from the apical cell border. Endocytosis of the colloid is dependent on the addition of membrane to the apical surface of the cell that occurs during exocytosis of thyroglobulin—ie, there is membrane recycling. (2) Electron-dense lysosomes migrate apically, meet the incoming colloid droplets, and fuse with them to produce phagolysosomes. The phagolysosomes continue their migration toward the basal aspect of the cell, during which time they become smaller and more dense as the lysosomal proteases hydrolyze the thyroglobulin. (3) T_4 and, to a much lesser degree, T_3, liberated from thyroglobulin by the proteolytic process, pass from the phagolysosome into the blood

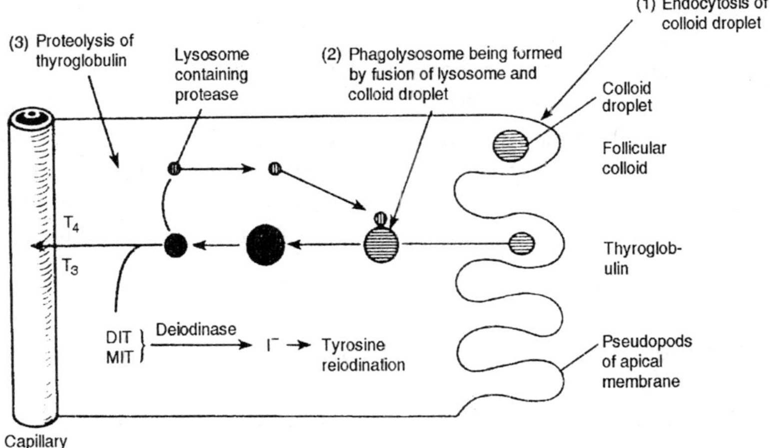

Fig. 7.10. Schematic representation of thyroid hormone secretion.

by a mechanism that is presently unknown, possibly diffusion. Most of the liberated iodotyrosines (MIT and DIT) are deiodinated by iodotyrosine deiodinase, releasing free iodide. This iodide is partly reutilized for thyroglobulin iodination, but some diffuses out into the circulation (iodide leak).

There is evidence that microtubules and microfilaments play an important role in these steps, presumably by regulating organelle movement. Thus, colchicine and cytochalasin B, which disrupt microtubular and microfilament function, respectively, inhibit thyroid hormone secretion.

Pharmacologic doses of iodide acutely reduce thyroid hormone secretion by an unknown mechanism, possibly by inhibiting thyroid cAMP generation. This slight effect on hormone secretion is clinically insignificant in the normal thyroid. In the hyperfunctioning thyroid, however, iodide has a marked inhibitory effect on hormone secretion that is clinically useful. Continued iodide administration usually inhibits thyroid hormone secretion for only a few weeks. The reason for the escape from this inhibitory effect is not known (and must not be confused with escape from the Wolff-Chaikoff block). Lithium acts in a manner similar to iodide, but lithium is rarely used clinically because it is more toxic.

In addition to the iodotyrosines, which are rapidly deiodinated by NADPH-dependent flavoprotein deiodinase within the thyroid, some T_4 released from thyroglobulin may also undergo intrathyroidal monodeiodination to T_3. The quantitative significance of this source of T_3 is uncertain. In the normally functioning thyroid gland, however, the secretion of T_3 (whether synthesized de novo or derived from T_4) is relatively small. The deiodination of iodotyrosines and the reutilization of this iodide are very important in the efficient utilization of iodide by the thyroid. This is because perhaps 80% of the iodide in thyroglobulin is in the form of iodotyrosines, which once secreted into the blood, are rapidly lost in the urine. There is evidence that intrathyroidal iodide released from iodotyrosines exists in a pool separate from iodide entering the thyroid cell directly from the extracellular fluid.

Regulation of Thyroid Function

Thyroid hormone synthesis and secretion are regulated by extrathyroidal (thyrotropin) and intrathyroidal (autoregulatory) mechanisms. Although thyrotropin (TSH) is the major modulator of thyroid activity, the importance of autoregulation is receiving increasing recognition.

TSH

TSH is a glycoprotein (molecular weight 28,000; 15% carbohydrate) secreted by specialized cells (thyrotrophs) in the anterior pituitary. TSH consists of 2 polypeptide chains. The α chain is nearly identical to the α chains of LH, FSH, and chorionic gonadotropin (hCG). TSH specificity is determined by the uniquely different β chain, which is critical for recognition of the TSH receptor. TSH biologic activity is present only when the 2 chains are combined. The half-life of TSH in plasma is very short (about an hour).

TSH secretion by the pituitary is modulated by thyroid hormone in a negative feedback regulatory mechanism. Recent evidence suggests that at the level of the pituitary, only T_3, produced locally by the monodeiodination of T_4, is able to inhibit TSH secretion. This is

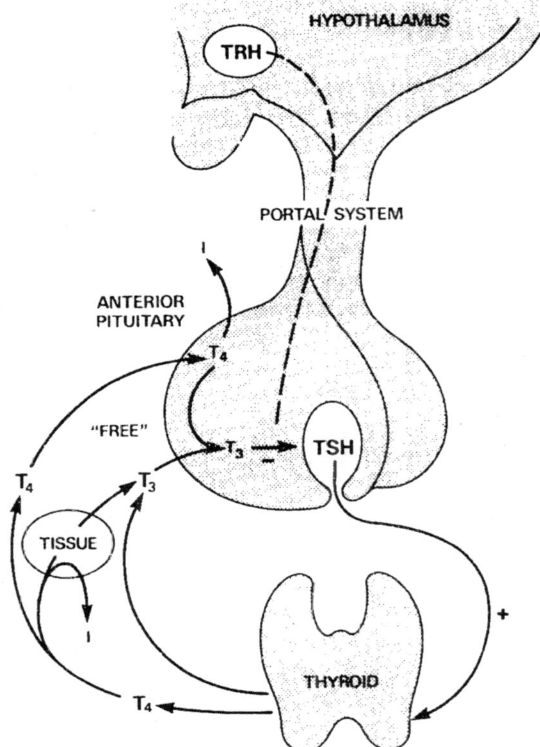

Fig. 7.11. The hypothalamic-hypophyseal-thyroid axis.

therefore the only site at which the monodeiodination of T_4 to T_3 has been shown to be a prerequisite for the expression of T_4 metabolic activity. In hypothyroidism, there is decreased suppression of TSH secretion, and serum TSH levels are higher than normal. The measurement by radioimmunoassay of high TSH levels in serum is very important clinically in the diagnosis of primary thyroid failure. A direct action of thyroid hormones on the hypothalamus has not been demonstrated.

The "thermostat" setting of the thyroid hormone-TSH feedback loop is modulated by a tripeptide, TSH-releasing hormone (TRH). This factor is present throughout the brain and indeed in other organs, but it is present at highest concentration in the hypothalamus. Hypothalamic production and release of TRH are controlled by poorly understood neural pathways from higher brain centers. TRH reaches the anterior pituitary from the hypothalamus via the hypothalamic-hypophyseal portal system. After binding to specific receptors on the thyrotroph, TRH increases TSH synthesis and secretion. While cAMP can stimulate TSH secretion by the thyrotroph, there is evidence that this agent may not mediate TRH-stimulated TSH release. The present concept is that TRH provides a tonic influence on the negative feedback of thyroid hormone on TSH secretion—ie, TRH is believed to regulate the sensitivity, or the set point, of the thyroid hormone-TSH negative feedback loop. Thyroid hormones, in turn, modulate anterior pituitary responsiveness to TRH by altering the number of TRH receptors on the thyrotrophs. Thus, in animals, administration of thyroid hormones decreases the number of TRH receptors, and in hypothyroidism the number of TRH receptors is increased.

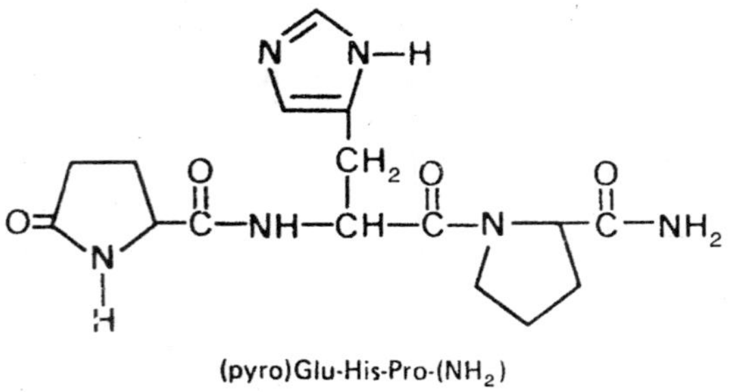

(pyro)Glu-His-Pro-(NH_2)

Fig. 7.12. Chemical structure of thyrotropin-releasing hormone (TRH).

Studies on the role of TRH in the regulation of TSH secretion in humans in vivo have been difficult because physiologically effective concentrations are attained only in the hypothalamo-hypophyseal portal system, and TRH reaching the systemic circulation, where it can be sampled and assayed, is greatly diluted and rapidly cleared with a half-life of a few minutes. In addition, TRH in the systemic circulation is derived from numerous tissues and presumably is represented only to a small degree by the TRH that reaches the systemic circulation after perfusing the pituitary. Because TRH is distributed throughout the central nervous system and because some studies have demonstrated an association between TRH and mood elevation, it is reasonable to speculate that this tripeptide may function as a neurotransmitter in addition to being a tonic regulator of TSH secretion.

Although it is not believed that acute pulses of TRH reach the pituitary and stimulate TSH secretion in vivo, the acute injection of a bolus of synthetic TRH (200-500 μg) rapidly increases TSH secretion. This response is of clinical value, particularly in hyperthyroidism. In this situation, despite TRH stimulation, TSH secretion remains suppressed by the high (even minimally elevated) thyroid hormone levels. However, other influences such as glucocorticoids and starvation may also reduce the TSH response to TRH stimulation. TRH stimulation is also valuable in the diagnosis of the very rare condition known as tertiary (hypothalamic) hypothyroidism, which is distinguished from

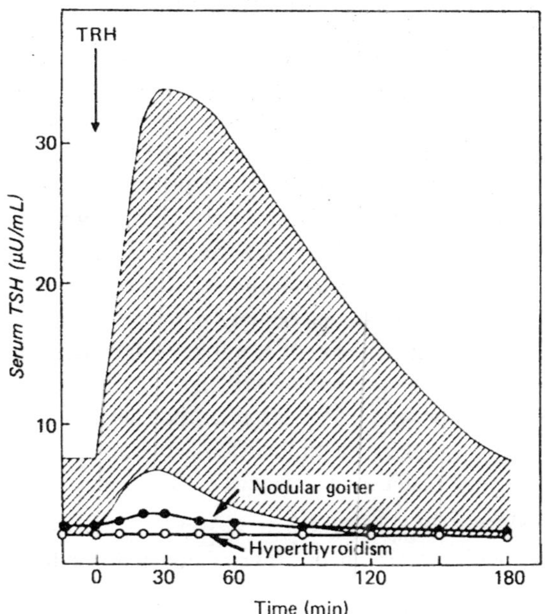

Fig. 7.13. Typical serum TSH responses to TRH.

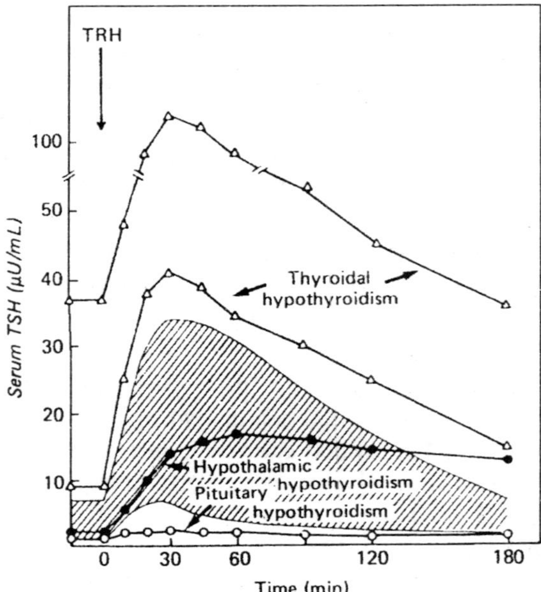

Fig. 7.14. Typical serum TSH response to TRH in patients with thyroid (primary), pituitary (secondary), and hypothalamic (tertiary) hypothyroidism.

secondary (pituitary) hypothyroidism by the presence of a TSH response to TRH in the former but not in the latter. The TRH test is of little clinical value in primary (thyroidal) hypothyroidism, where the high serum TSH level even before TRH injection is diagnostic and the exaggerated TSH response provides little additional information.

Acute injection of TRH also stimulates prolactin release by the pituitary, but it is not clear whether this is physiologically important. In some patients with hypothyroidism, however, serum prolactin levels may be elevated. Dopaminergic stimuli such as the administration of levodopa reduce both the prolactin and TSH response to TRH stimulation. In addition, levodopa acutely depresses the elevated TSH levels in primary hypothyroidism. It is unlikely, however, that TRH is the sole releasing factor for both TSH and prolactin, because their secretion may be dissociated under certain circumstances. For example, the nocturnal elevation in prolactin and the acute prolactin response to suckling are not associated with changes in TSH secretion.

Actions of TSH on the Thyroid

TSH has many effects on the thyroid the net result of which is increased thyroid hormone secretion.

1. Most TSH actions are produced by binding to specific thyroid plasma membrane receptors with the subsequent stimulation of adenylate cyclase activity and cAMP generation.
2. TSH affects the iodide transport mechanism in a biphasic manner. There is an initial acute (4-hour) decrease in the iodide thyroid/serum (T:S) ratio. This effect is also produced in vitro by cAMP and reflects an increased rate of I^- efflux. After 4 hours of continued TSH stimulation, the iodide T:S ratio increases as the iodide transport mechanism is enhanced in a process dependent on new protein synthesis.
3. TSH (and cAMP) stimulate iodide organification (incorporation of iodide into thyroglobulin) primarily by increasing H_2O_2 generation. Exocytosis of thyroglobulin into the follicular lumen is also increased.
4. Acute TSH stimulation (1-2 minutes) increases pseudopod formation at the apical cell border, followed by endocytosis of colloid, phagolysosome formation, and the subsequent secretion of thyroid hormones. As mentioned above, these effects are blocked by colchicine and cytochalasin, which disrupt the cytoskeletal system.
5. Chronic TSH stimulation increases thyroid transcriptional and translational activity and ultimately produces hyperplasia and goiter.
6. Many other effects of TSH on thyroid intermediary metabolism have been described, including increased glucose oxidation (primarily via the pentose phosphate pathway), phospholipid turnover, and increased precursor uptake into thyroid cells. Not all of these effects are produced by cAMP.

Other Thyroid Stimulators

Other stimulators of cAMP generation, particularly epinephrine and prostaglandins, can increase thyroid hormone secretion under experimental conditions. These agents mimic many (not all) of the effects of TSH. In the case of catecholamines, autonomic nerve endings have been demonstrated adjacent to the thyroid follicular cells, and it is conceivable that they may play a role in the regulation of thyroid function. At present, however, their role remains speculative and is most likely a minor one if it exists at all.

Thyroid Autoregulation

The thyroid is able to regulate its uptake of iodide and thyroid hormone synthesis by intrathyroidal mechanisms independent of TSH—ie, these regulatory mechanisms occur even in hypophysectomized animals.

Wolff-Chaikoff block

When increasing amounts of I^- are given to experimental animals, inhibition of iodide organification occurs at a critical level of intrathyroidal inorganic I^-. Thyroglobulin iodination and thyroid hormone synthesis subsequently decrease. This is known as the Wolff-Chaikoff block.

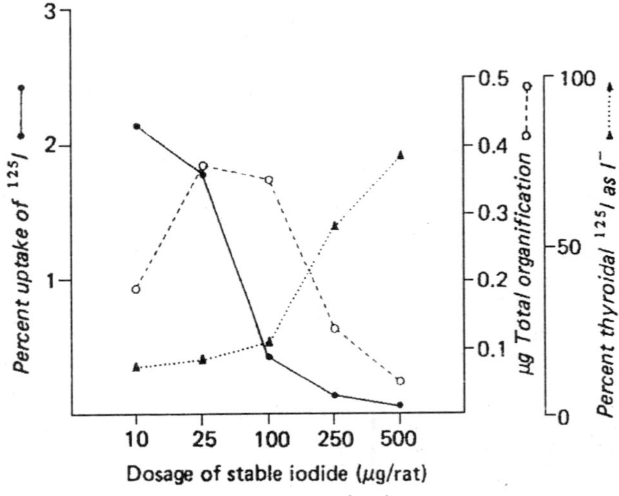

Fig. 7.15. The Wolff-Chaikoff block.

The normal thyroid gland escapes from the Wolff-Chaikoff block—and hypothyroidism does not ensue—because of intrathyroidal feedback inhibition of the iodide transport mechanism by an as yet unidentified organic ioaine intermediate: ie, there is a reduction in the ability of the thyroid to trap iodide and to maintain a high T:S ratio. Because of diminished active iodide transport, the intrathyroidal inorganic I^- concentration then declines and the "brakes" come off the organification block. A new equilibrium now exists in which thyroid hormone is synthesized at the rate maintained before imposition of the excess iodide load, though the fractional thyroid uptake of iodide from the extracellular fluid is reduced. Conversely, in iodide insufficiency, the iodide transport mechanism becomes more active (independent of TSH) and there is a greater fractional uptake of iodide from the plasma perfusing the thyroid gland. This is reflected in an increased radioactive iodine uptake. In some thyroid diseases (eg, Hashimoto's thyroiditis), escape from the Wolff-Chaikoff block does not occur, and hypothyroidism (iodide myxedema) ensues.

Sensitivity to TSH

The thyroid is able lo "sense" its organic iodide content and alter its sensitivity to TSH stimulation. For example, iodide-deficient animals injected with TSH develop bigger goiters than do iodide-sufficient animals injected with the same amount of TSH. An important set point for this, as well as for the iodide transport mechanism, may be at the level of adenylate cyclase. Thus, an organic iodide intermediate also inhibits the cAMP response to TSH stimulation. Whether or not this inhibitor is the same as the inhibitor of the iodide transport mechanism remains to be determined.

Thyroid hormone secretion

The Wolff-Chaikoff block must be distinguished from the rapid and transient decrease in thyroid hormone secretion induced by iodide, especially in hyperthyroidism. Although the iodide organification block occurs rapidly, its clinical effects are more delayed in susceptible individuals, because follicular thyroid hormone stores must first be depleted before thyroid hormone secretion declines. On the other hand, a block in hormone secretion can lower serum hormone levels even with full follicular stores of thyroid hormone.

Other examples of autoregulation

During periods of iodide insufficiency, the ratio of T_3 to T_4 secreted by the thyroid is increased. Since T_3 is more potent than T_4, the thyroid is therefore able to utilize available iodide more efficiently when the supply is limited. Poorly iodinated thyroglobulin is also more susceptible to proteolysis into thyroid hormones than is heavily iodinated thyroglobulin.

Thyroid Hormones in Plasma

Plasma Binding Proteins

Thyroid hormones (T_4 and T_3) in plasma are largely bound to protein. Only about 0.04% of the T_4 and 0.4% of the T_3 circulate "free" in the unbound state. The present concept is that only the free hormones enter cells, produce their biologic effects, and are in turn metabolized. Only the free hormone regulates the pituitary feedback mechanism. A dynamic equilibrium exists between the plasma and intracellular free hormone pools. Protein-bound thyroid hormones are therefore a very large reservoir that is slowly drawn upon as the free hormone dissociates from the binding proteins and enters the cells. Thyroid secretion is precisely regulated to replenish the metabolized thyroid hormones. This maintains a constant level of free hormone.

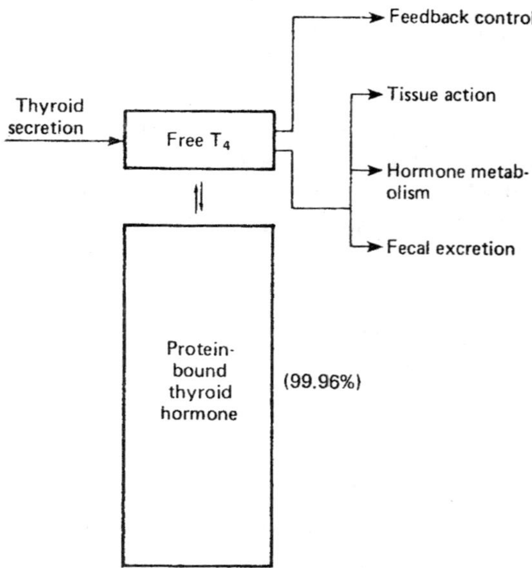

Fig. 7.16. Representation of free T_4 (and free T_3) as the biologically active hormone at the level of the pituitary and the peripheral tissues.

Very little thyroid hormone is lost in the urine, because protein binding minimizes glomerular filtration. Metabolites of T_4 and T_3, on the other hand, are very poorly bound to plasma proteins and are therefore rapidly excreted in the urine.

There are 3 major thyroid hormone-binding proteins in plasma. The hormone-carrying capacity of these proteins can be determined by adding radioactively labeled hormone to serum and subjecting it to electrophoresis.

Thyroid hormone-binding globulin (TBG)

TBG is a monomeric glycoprotein. It is the thyroid hormone-binding protein at lowest concentration in plasma (1-2 mg/dL). Each molecule has a single binding site for T_4 or T_3. Despite its very low concentration, it is the protein with the highest affinity for thyroid hormones ($T_4 > T_3$) and therefore carries most (70%) of the bound thyroid hormones. When fully saturated, it can carry about 20 μg of T_4 per deciliter.

Thyroxine-binding prealbumin (TBPA)

TBPA is present at intermediate concentration in plasma (25 mg/dL). TBPA binds essentially no T_3. Even though its concentration in plasma is 20 times that of TBG, TBPA binds much less T_4 than TBG,

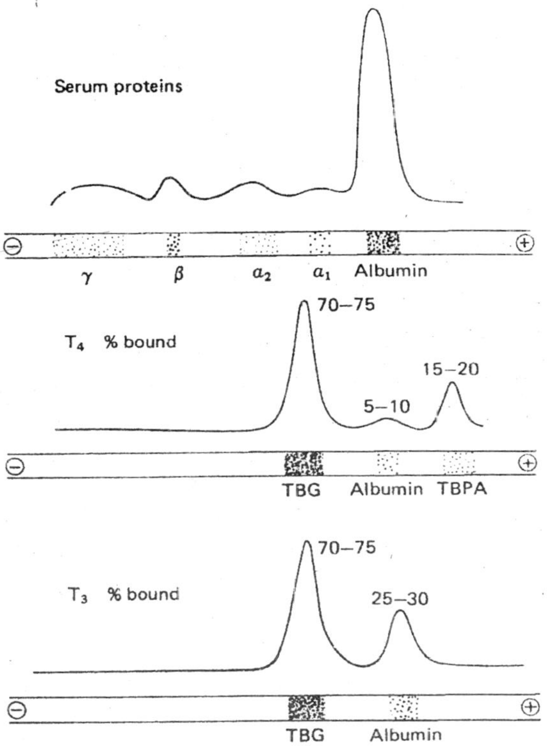

Fig. 7.17. Diagrammatic representation of the distribution of radioactive T_4 and T_3 among serum thyroid hormone-binding proteins. Top: Paper electrophoretic pattern of serum proteins. Middle: Radioactive T_4 was added to serum and was then subjected to paper electrophoresis. Bottom: Radioactive T_3 was added to serum and subjected to paper electrophoresis.

because it has a much lower affinity for T_4. Thus, about 20% of T_4 plasma is bound to TBPA—in contrast to approximately 70% of T_4 bound to TBG.

Albumin

Albumin has the lowest affinity for T_4 and T_3. However, it binds significant amounts of thyroid hormone (approximately 10% of T_4 and 30% of T_3), because it is present in plasma at very high concentration (about 3500 mg/dL).

Kinetics of thyroid hormone interaction with plasma binding proteins

T_4 and T_3 bind reversibly with each binding protein according to the *law of mass action*. For example, in the case of T_4,

$$(T_4) + (TBG) = (TBG\text{-}T_4) \quad ...(1)$$

where (T_4) represents free (unbound) T_4, (TBG) is TBG not containing T_4, and (TBG-T_4) is T_4 bound to TBG (all in molar concentrations). At equilibrium, the rate of association of T_4 and TBG to form TBG-T_4 is the same as the dissociation of TBG-T_4 into its individual components. Thus,

$$kT_4 = \frac{(TBG - T_4)}{(T_4)\,(TBG)} \qquad \ldots(2)$$

where kT_4, is the association constant for the interaction of T_4 with TBG.

From equation (2),

$$(T_4) = \frac{(TBG - T_4)}{kT_4\ (TBG)} \qquad \ldots(3)$$

A similar relationship exists for each of the 3 major binding proteins, with their different affinities for T_4 (or T_3), as well as their different absolute concentrations. The free T_4 concentration in serum containing all of these binding proteins is the sum of the values in the numerator (equation 3) for each protein divided by the sum of the respective denominators.

It can therefore be seen from equation (3) that an increase in the concentration of a particular binding protein will initially result in a

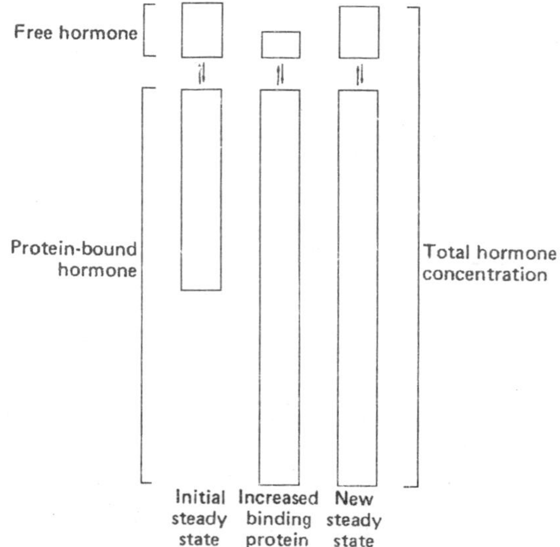

Fig. 7.18. The effect of an increase in thyroid hormone-binding protein concentration on the free and protein-bound hormone concentrations.

decrease in the free T_4 concentration. In the presence of a normal hypothalamic-pituitary-thyroid axis, there will be a compensatory increase in thyroid hormone secretion until the serum free T_4 level returns to normal. Under these conditions, the normal free T_4 concentration is now associated with an increase in the total (protein-bound) T_4. Daily free T_4 production and clearance rates are the same in both the "initial" and "new" conditions despite the fact that total plasma T_4 concentrations are different.

Because of the high affinity of the binding proteins for thyroid hormones, the concentrations of free T_4 and T_3 in plasma are negligible compared to the concentrations of protein-bound hormones. The former values can therefore be ignored in expressing the total thyroid hormone concentration in plasma. If the thyroid hormone concentration increases to the point of saturation for TBG (approximately 20 μg T_4 per deciliter), as may occur in hyperthyroidism, any further increase in total serum T_4 concentration is accompanied by a much larger increase in free T_4.

In the converse situation, if the concentration of thyroid hormone-binding protein in plasma is decreased, or if hormonal binding to the proteins is inhibited by, for example, a pharmacologic agent, there is an initial transient increase in the serum free T_4 concentration. This suppresses TSH secretion, and thyroid hormone secretion is consequently reduced. With continued utilization of thyroid hormone by the tissues, the free thyroid hormone level decreases to the normal "euthyroid" level. At equilibrium, the normal free hormone level is now associated with a decreased total (protein-bound) hormone concentration.

The binding of thyroid hormones to plasma binding proteins may vary considerably. Low or high TBG levels are uncommon genetic traits that can produce abnormal levels in total hormone concentrations, with the free (metabolically active) hormone levels remaining normal. In addition, a variety of diseases and pharmacologic agents can either decrease the amount of binding protein present in plasma or influence the binding of thyroid hormones to a normal concentration of binding proteins. Because of the high affinity of TBG for T_4 and T_3, alterations in its concentration have a much greater effect on serum total hormone concentrations than do alterations in TBPA or albumin concentrations.

Thyroid Hormone Metabolism

T_4 is the major secretory product of the normal thyroid. The major pathway of T_4 metabolism is via the progressive deiodination of the molecule. The initial deiodination of T_4 may occur in the outer

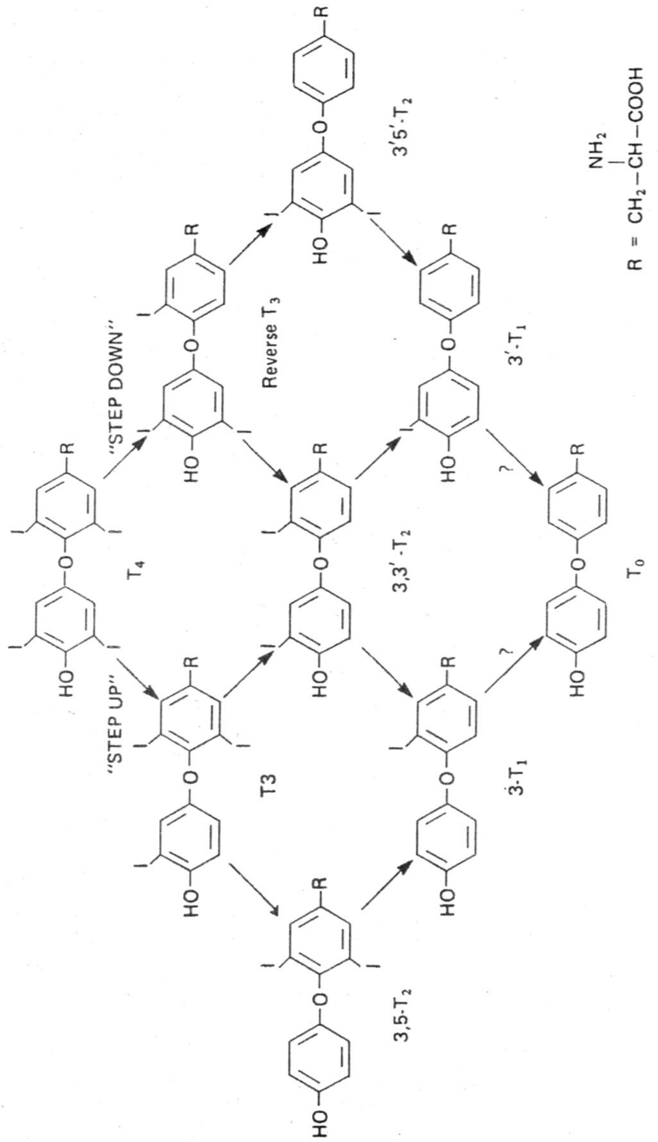

Fig. 7.19. The deiodinative pathway of thyroxine metabolism. The monodeiodination of T_4 to T_3 represents a "step up" in biologic potency, whereas the monodeiodination of T_4 to reverse T_3 has the opposite effect. Further deiodination of T_3 essentially abolishes hormonal activity.

ring, producing T_3 (3,3´,5-T_3); or in the inner ring, producing reverse T_3 (rT_3; 3,3´,5´-T_3). Less than 20% of total T_3 is produced in the thyroid. The remaining 80-90% is derived from outer (phenolic) ring monodeiodination of T_4 in the peripheral tissues. T_4 and T_3 in the plasma, are metabolized by the peripheral tissues and subsequently

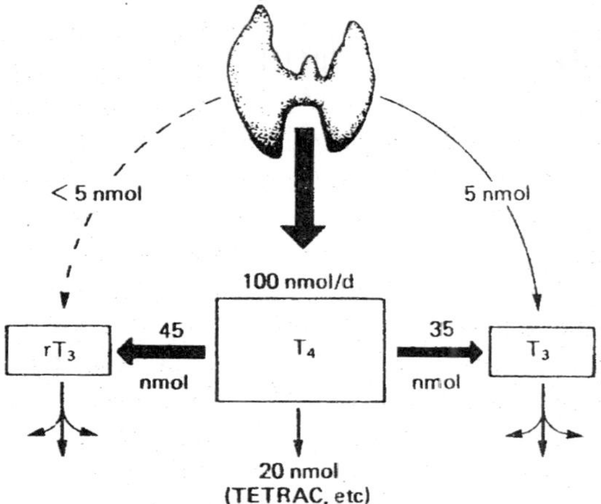

Fig. 7.20. Major pathways of thyroxine metabolism in normal adult humans.

excreted. Although the catabolic products of T_4 are generally less biologically active than the parent compound, some metabolites do have biologic activity and, as in the case of T_3, even exceed the biologic potency of T_4. It should be noted that data on the bioactivity of compounds other than T_4 and T_3 are based on difficult and inaccurate assays and should be regarded merely as approximations. Almost all (95%) of the reverse T_3 (rT_3) in the plasma is derived from the inner ring monodeiodination of T_4 in the peripheral tissues. Both T_3 and rT_3 are then deiodinated further to the diiodothyronines (T_2) and then to monoiodothyronines. Reverse T_3 and the 3 different T_2 metabolites are essentially biologically inactive and are cleared rapidly from the plasma.

Because conversion of T_4 to T_3 represents a "step up" in biologic activity whereas conversion of T_4 to rT_3 has the opposite effect, the conversion of T_4 to T_3 or rT_3 by inner or outer ring iodothyronine deiodinase is a pivotal regulatory step in determining thyroid hormone biologic activity. It has been proposed that T_4 is a biologically inactive prohormone, and that its apparent biologic activity is dependent upon intracellular conversion to T_3. However, other than in the pituitary, all evidence indicates that T_4 itself has intrinsic biologic activity.

The kinetics of T_4, T_3, and rT_3 production, distribution, and clearance, which indicates typical values that may be seen in a normal human. It can be seen that because of its greater binding affinity to plasma proteins, chiefly TBG, the distribution of T_4 is largely restricted to the intravascular volume; ie, T_4 has the smallest distribution volume and the largest pool size. Because of the large plasma pool of T_4 relative to its rate of utilization, the plasma T_4 reservoir turns over relatively slowly. T_3 and, to a greater degree, rT_3 are more loosely bound to plasma proteins and are therefore distributed in a larger volume and cleared more rapidly.

The tissues that concentrate the most thyroid hormone are liver, kidney, and muscle. Brain, spleen, and gonads take up little thyroid hormone. This variation in uptake by different tissues parallels the metabolic effects of T_4 on oxygen consumption in vitro.

It is uncertain how thyroid hormones enter peripheral tissue cells. It was at one time thought that this was a passive diffusion process, but the possibility that there may be specific thyroid hormone receptors on the cell surface and active transport of thyroid hormone into cells is under active investigation. Once within the cell, thyroid hormones either remain free in the cytoplasm or are associated with low-affinity cytosol receptors or high-affinity nuclear and mitochondrial receptors.

In addition to the deiodinative pathways of T_4 metabolism, minor metabolic pathways of uncertain importance include the following:

1. Oxidative deamination of the iodothyronines into their acetic acid derivatives, eg, tetraiodothyroacetic acid (TETRAC) and triiodothyroacetic acid (TRIAC). These compounds retain some biologic activity, but their relative physiologic importance is uncertain.
2. Phenolic conjugation into glucuronide or sulfate derivatives. The former occurs mainly in the liver and the latter mainly in the kidney. Glucuronide conjugates of T_4 are excreted into the bile and thence in feces. In humans, there is little enterohepatic circulation of thyroid hormone conjugates.
3. Decarboxylation of thyroxine to thyroxamine.
4. Cleavage of the ether link between the 2 phenol rings of the iodothyronines.

As mentioned above, 80-90% of T_3 production is from the monodeiodination of T_4 to T_3. This conversion of T_4 to T_3 utilizes approximately 30-40% of the secreted T_4. Virtually all rT_3 (95%) is produced by the monodeiodination of T_4 and this represents another 40% of the T_4 secreted by the thyroid. The remaining 20% of secreted T_4 is excreted in the feces or urine, either in the free form or as conjugates.

A variety of conditions and pharmacologic agents reduce the rate of T_4 monodeiodination to T_3. This may lead to decreased serum T_3 concentrations. The clinical status of a patient is therefore very important in interpreting the significance of a low serum T_3 level. In acute illness or starvation, the low serum T_3 value is usually associated with an elevated serum rT_3 concentration. This reflects primarily decreased rT_3 clearance rather than increased rT_3 production. While rT_3 clearly has little or no agonist activity, there is evidence that it may indeed be a counterhormone in reducing the metabolic activity of more active thyroid hormones.

Thyroid Hormone Actions

Administration of T_4 or T_3—or the pathologic or iatrogenic absence of these hormones—produces general effects on intermediary metabolism and has particular effects on specific organ systems.

Fetal Development

Thyroid hormones are critically important in fetal development, particularly of the neural and skeletal systems. Thus, intrauterine hypothyroidism leads to cretinism (mental retardation and dwarfism). Maternal thyroid hormones do not cross the placenta in sufficient quantity to maintain fetal euthyroidism, and the fetus is therefore dependent upon hormones synthesized by its own thyroid gland (from about 11 weeks' gestation).

Oxygen Consumption and Heat Production

The basal metabolic rate (O_2 consumption, at rest, by the whole animal) increases in hyperthyroidism and decreases in hypothyroidism.

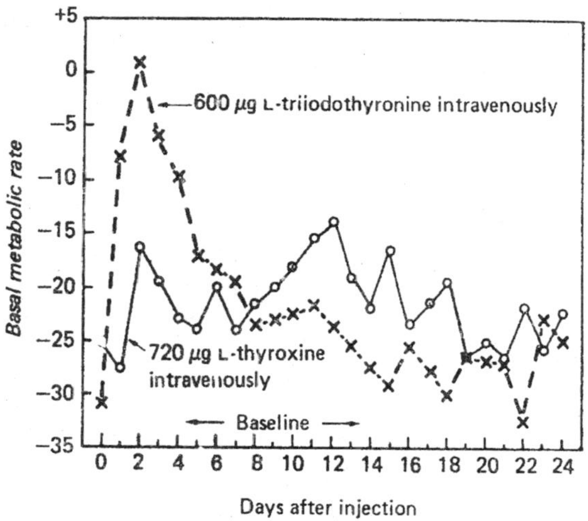

Fig. 7.21. Basal metabolic rates after injection of equimolar amounts of T_3 and T_4 in a hypothyroid individual.

Severe hyperthyroidism is occasionally associated with fever and severe hypothyroidism with hypothermia. Postnatally, thyroid hormones increase O_2 consumption in all tissues except the brain, spleen, and testis. Compared to the effects of TSH on thyroid hormone secretion, most thyroid hormone actions on peripheral tissues are induced relatively slowly over a period of hours or days. For example, T_4 administration to hypothyroid patients does not increase oxygen consumption until 24-48 hours after administration. T_3 acts more rapidly than T_4 but still over a period of hours.

Cardiovascular System

Thyroid hormones have marked chronotropic and inotropic effects on the heart. Low cardiac output with bradycardia and slow myocardial contraction nd relaxation are characteristic of hypothyroidism. The reverse occurs in hyperthyroidism. The force of contraction in hypothyroidism is normal.

Sympathetic Effects

Many thyroid hormone effects, particularly on the cardiovascular system, are similar to those induced by catecholamines. Hyperthyroid patients are more sensitive to catecholamines. Thyroid hormones increase the number of catecholamine receptors in heart muscle cells. Hepatocyte receptors are unaffected. Thyroid hormones may also amplify catecholamine action at a postreceptor site. Beta-adrenergic inhibition reverses some features of clinical hyperthyroidism, such as eyelid retraction and tachycardia. However, other thyroid hormone actions—-eg, effect on O_2 consumption—are not prevented by beta-adrenergic blockade. It is clear, therefore, that thyroid hormone action is distinct from catecholamine action. Their exact interrelationship remains to be elucidated.

Pulmonary Effects

Thyroid hormones are necessary for normal hypoxic and hypercapnic drive to the respiratory centers. Hypoventilation with hypoxia and hypercapnia is a consequence of severe hypothyroidism and may require artificial ventilation.

Hematopoiesis

Thyroid hormones increase erythropoiesis, possibly because of increased O_2 utilization by tissues leading to increased erythropoietin production. Thyroid hormones also increase 2,3-diphosphoglycerate concentrations in erythrocytes, allowing increased O_2 dissociation from hemoglobin and thereby increasing O_2 availability to the tissues. In hypothyroidism, the reverse occurs, and decreased O_2 consumption may lead to "anemia" (which, however, is physiologic).

Endocrine System

Thyroid hormones have a general effect in increasing the metabolism and clearance of various hormones and pharmacologic agents. For example, steroid hormone clearances are increased, leading to compensatory increases in production rates. Thus, administration of thyroid hormones will increase cortisol production and clearance, but the plasma cortisol concentration remains unchanged. Serum prolactin levels are increased in about 40% of patients with primary hypothyroidism. When present, this abnormality is corrected by treatment with thyroid hormone. Insulin requirements in diabetics are frequently increased in hyperthyroidism. The growth hormone response to stimuli such as hypoglycemia is reduced in primary hypothyroidism. Thyroid hormones are necessary for normal LH and FSH secretion. In hypothyroidism, anovulation and menstrual disturbances—particularly menorrhagia—may occur. Decreased renal free water clearance in hypothyroidism may be secondary to enhanced vasopressin activity but is more apt to be related to altered intrarenal hemodynamics. Finally, PTH action may be diminished in hypothyroidism.

Musculoskeletal System

Thyroid hormones have a potent stimulatory effect on bone turnover, increasing both bone formation and resorption. This is associated with increased urinary hydroxyproline excretion. Hypercalcemia is occasionally observed in severe hyperthyroidism. Thyroid hormones increase the rate of muscle relaxation as measured by elicitation of the deep tendon reflexes. This is of clinical value in diagnosing hypothyroidism.

Thyroid Hormone Action at the Molecular Level

Actions of thyroid hormones have been described at different intracellular sites. There is presently much discussion regarding their relative importance.

RNA Transcription

There is considerable evidence that tree T_4 and T_3 enter tissue cells and bind to specific receptors in nuclear acidic proteins associated with chromatin. This is followed by increased RNA transcriptional activity. The physiologic importance of these nuclear binding sites is suggested by their high affinity constant, which is compatible with the prevailing plasma levels of free thyroid hormones, in addition, there is a good correlation between the binding of various analogs of thyroid hormones to nuclear receptors in liver and their metabolic potency. The exact mechanism by which thyroid hormone binding to nuclear protein receptors leads to increased transcriptional activity is under active investigation.

Protein Translation

T_4 and T_3 increase detectable new protein synthesis in vitro within 5 hours. This is sooner than an effect on transcription can be observed. It is therefore difficult to explain this effect on new mRNA synthesis, and thyroid hormones are therefore believed to alter protein synthesis at a posttranscriptional site as well. Correlation of the induction by thyroid hormones of different enzymes such as mitochondrial malic enzyme and cytosolic α-glycerophosphate dehydrogenase with their respective mRNAs provides further evidence for posttranscriptional effects of thyroid hormones.

Cell Membrane

Thyroid hormones markedly stimulate plasma membrane Na^+-K^+ ATPase activity, which is coupled to Na^+ and K^+ transport across the plasma membrane. This increases ATP utilization. Since a large portion of the O_2 consumption by the entire organism is for maintenance of this transport system, it has been suggested that the effect of thyroid hormone in increasing the basal metabolic rate is related to this effect.

Thyroid hormones have also recently been found to have an acute (within minutes) effect on the plasma membrane of cells such as lymphocytes in that they increase the intracellular transport of glucose and amino acids. The physiologic importance of this phenomenon has not been established.

Mitochondria

Although it was initially thought that thyroid hormones increased oxygen consumption by uncoupling mitochondrial oxidative phosphorylation, leading to inefficient energy utilization, this is now thought to be unlikely. It is possible that ATP synthesis may he increased as a secondary phenomenon—eg, in response to increased ATP utilization in association with increased Na^+ and K^+ transport activity. Nevertheless, mitochondrial receptors of high affinity for thyroid hormones have recently been described, and their importance in mitochondrial function is being investigated.

Effects on Intermediary Metabolism

Thyroid hormones alter the metabolism of carbohydrates, fats, and proteins. It is likely, however, that these effects are not primary but secondary to the mechanisms described above. Cholesterol synthesis and degradation are both increased by thyroid hormones. Since the latter effect predominates, the serum cholesterol level declines in thyroid overactivity and vice versa. Thyroid hormone-induced effects on cholesterol metabolism may be secondary to increased beta-lipoprotein degradation. This may also account for the alteration in serum carotene that parallels the alteration in serum cholesterol. Lipolysis is also increased by thyroid hormones, possibly by enhancing the response to catecholamine stimulation. Glycogenolysis and gluconeogenesis are increased by thyroid hormones, presumably in relation to regeneration of ATP consumed—eg, by increased Na^+-K^+ ATPase activity. Thyroid hormones increase insulin requirements in diabetics. The reasons for this are unclear and may be related to increased insulin degradation or clearance rather than to alterations in carbohydrate metabolism.

Thyroid Function during Growth and Development

Fetal Development

Thyroglobulin synthesis begins during the fifth week of intrauterine life, and the iodide transport mechanism becomes active at approximately 12 weeks. During pregnancy, the fetal thyroid is under the control of the fetal hypothalamus and pituitary. Maternal thyroid hormones cross the placenta only to a slight extent: thus, athyreotic fetuses have detectable but low (hypothyroid) thyroid hormone concentrations.

At maturity, the human fetus has a slightly in- creased serum TSH and free thyroxine concentration. However, the serum T_3 level is very low—in the range noted in adults during starvation and acute illness. Similarly, the rT_3 concentration is elevated. Immediately after birth, there is a transient increase in TSH secretion, and serum T_3, T_4, and rT_3 levels revert to nearly normal levels within the first few days and approach normal adult levels during the first year of life.

Thyroid Function in Pregnancy

Serum TBG levels increase approximately 2-fold during pregnancy. A new equilibrium is reached in which the total thyroid hormone concentrations are twice normal but free hormone levels remain normal. Radioactive iodine uptake increases during pregnancy, and the thyroid may increase slightly in size, because of increased renal clearance of iodide and perhaps because of increased iodide utilization by the fetus. The placenta produces a glycoprotein hormone with very weak thyrotropic activity. It is unclear whether this is human chorionic gonadotropin (hCG) itself or another closely related compound that is extracted from the placenta together with hCG. Recent evidence suggests the latter. In normal pregnancy, this substance has little or no effect on thyroid function.

Changes in Thyroid Function With Aging

Although serum T_3 levels are reported to decrease slightly with aging, this observation has been questioned, because elderly persons are more susceptible to a variety of illnesses that may reduce serum T_3 production from T_4 and therefore reduce the average T_3 level in the group. Irrespective of whether an age-related decrease in T_3 occurs or not, thyroid function in old people appears to be normal as shown by the TSH response to TRH stimulation.

Tests of Thyroid Function

Thyroid function tests may be classified as (1) those that measure concentrations of hormones or biologically inactive products secreted by the thyroid gland; (2) those that directly examine the function of the thyroid gland: (3) those that test the effects of thyroid hormones on peripheral tissues: (4) those that evaluate the hypothalamic-pituitary-thyroid axis: and (5) those that identify and measure pathologic substances in blood not usually present in normal subjects.

Measurement in Blood of Products Secreted by the Thyroid

Measurement of T_4

T_4 is measured by competitive binding assay or by radioimmunoassay. The former test is based on the very high affinity of TBG for T_4. For the latter, an antibody is raised in rabbits to T_4 (conjugated to a larger protein to be more antigenic). In both instances, the conditions of the assay are such that T_4 is dissociated from endogenous TBG, which therefore does not interfere in the assay. The unknown serum is compared to a series of standards of known T_4 concentration in their ability to competitively inhibit the binding of a tracer amount of radioactive T_4 to TBG or to anti-T_4 antibody.

Under some circumstances, the total T_4 concentration in serum does not accurately reflect metabolic status. As described previously, thyroid hormone-binding protein concentrations may be altered in a variety of conditions. This leads to altered total T_4 levels but not the metabolically important free T_4 concentrations. Certain drugs—phenytoin and salicylates are the most common offenders—interfere with the binding of T_4 to serum binding proteins. For this reason, it is of clinical value to combine the estimation of the total serum T_4 with an indirect measurement of the free T_4 concentration.

Free T_4 Index

Direct measurement by dialysis of serum free T_4 is a difficult and laborious procedure because of its very low concentration. Of more practical value for routine determinations, an approximation of the true free T_4 concentration may be obtained and expressed as a free T_4 index (FT_4I). For this purpose, an estimate is made of the binding capacity of the patient's serum for thyroid hormones. Radiolabeled T_3 or T_4 (usually T_3) is first added to the serum to be tested and allowed to equilibrate with unlabeled endogenous hormone. An aliquot is then added to either a resin or Sephadex—substances that bind free thyroid hormone with high affinity and therefore compete with serum binding proteins for the free hormone. More T_3 than T_4 is bound to resin because T_3 binds with lower affinity to endogenous binding proteins. The assay using radiolabeled T_3 is therefore more practical than the one using T_4. The resin is then rinsed and counted for radioactivity to determine what percentage of the total radioactivity has been adsorbed. The percentage uptake of total radioactive T_3 by

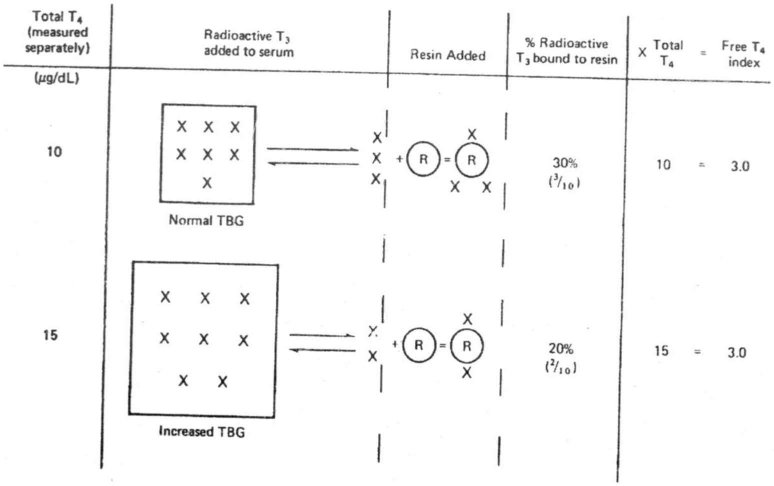

Total T_4 (measured separately) (μg/dL)	Radioactive T_3 added to serum	Resin Added	% Radioactive T_3 bound to resin	× Total T_4	= Free T_4 index
10	Normal TBG		30% (3/10)	10	3.0
15	Increased TBG		20% (2/10)	15	3.0

Fig. 7.22. Use of the free T_4 index (FT_4I) to estimate the free T_4 (or free T_3) concentration in serum independently of changes in the thyroid hormone-binding protein concentration.

the resin will be greater when there is less binding of the radioactive T_3 to serum binding proteins. Conversely, if serum binding proteins are increased, as in pregnancy, the percentage uptake of radioactive T_3 by the resin will decrease.

Multiplication of the resin or Sephadex uptake value by the total T_4 concentration provides an index of the effective free T_4 concentration independently of changes in hormone binding capacity in serum. It is important to emphasize that this is not a true free T_4 determination but merely an index of this measurement. It has been demonstrated by direct measurement of free T_4 levels by dialysis that there is reasonable proportionality between the true free T_4 level and the free T_4 index, thereby establishing the clinical usefulness of the latter. It may seem confusing that the free T_4 index (FT_4I) is measured with radioactive T_3 ("T_3 resin uptake test"; RT_3U). However, the test is an assessment of the capacity for serum proteins to bind thyroid hormones, and it makes no difference whether T_3 or T_4 is used. T_3 is more useful, because it is not as tightly bound to serum proteins as T_4, and the resin is better able to compete for available free hormone.

In most cases, the combination of the total T_4 and FT_4I is sufficient to establish the thyroid metabolic status of the patient. Under some circumstances, however, even when alterations in thyroid hormone-binding proteins are taken into account by the FT_4I, the value may be misleading and obscure the correct diagnosis. If there is suspicion that the T_4 and FT_4I results are clinically inappropriate, additional tests are necessary. Thus, the FT_4I may be inappropriately low in patients ingesting excessive amounts of T_3 (thyrotoxicosis factitia) or may be normal when hyperthyroidism is associated with relative overproduction of T_3 but not T_4 (T_3 thyrotoxicosis). Phenytoin is a unique drug in that it may lower the FT_4I as well as the total T_4 even though patients remain clinically euthyroid by all other available criteria. The reason for this is unclear. A possible explanation is that intracellular T_4 (or T_3) levels are normal despite depressed serum free T_4 levels.

Occasionally, the FT_4I is inappropriately high. An example of this is the very rare condition of peripheral resistance to thyroid hormones. In addition, during acute illness, euthyroid patients may rarely present with an elevated FT_4I; however, the total T_3 concentration is subnormal. Occasionally, in acute or chronic illness, the resin uptake test for the FT_4I is artificially interfered with by unknown serum factors, and an inappropriately low resin uptake is measured. This results in an artifactually low FT_4I, and the patient may be wrongly diagnosed as being hypothyroid. In these patients, measurement of the true free T_4 by dialysis reveals a normal or even high level. The diagnosis should therefore be reassessed after recovery from the illness. If waiting is not prudent, primary hypothyroidism can be excluded by a serum TSH determination.

Measurement of T_3

The T_3 concentration in serum is generally measured by radioimmunoassay (RIA). This should not be confused with the resin T_3 uptake test (RT_3U), which does not measure the serum T_3 level. The competitive binding assay with TBG is more difficult for T_3 than for T_4, because TBG also binds T_4, which is present in serum at a higher concentration than T_3 and which binds with higher affinity lo TBG. Measurement of T_3 is of more value in the diagnosis of hyperthyroidism than hypothyroidism, because in hyperthyroidism the increase in T_3 secretion is proportionately greater than the increase in T_4 secretion. This may be manifested occasionally as "T_3 thyrotoxicosis," in which an elevated serum T_3 level is associated with a normal T_4 level. In contrast, in mild hypothyroidism, the serum T_3 level does not fall to the same degree as does the T_4, because TSH stimulation increases the relative secretion of T_3. On the rare occasion when ill patients are found to have a slightly elevated free T_4 index, measurement of the total T_3 level is of great value in excluding hyperthyroidism. In a wide variety of acute and chronic illnesses, the serum total T_3 level is suppressed because of decreased conversion of T_4 to T_3.

Free T_3 Index

The total T_3 concentration is influenced by alterations in the concentration of serum binding proteins in the same way as is the serum total T_4 level. Thus, just as the high total T_4 in a patient taking estrogens for contraception is difficult to interpret without determination of the free T_4 index, so must a high total T_3 (RIA) level under these circumstances be interpreted with caution. Use of the free T_3 index is increasing. The test is performed as described above for free T_4 index. The product of the T_3 (RIA) and the RT_3U allows the calculation of a free T_3 index (FT_3I) that is an indicator of the free T_3 level in serum. It is important to mention again that the T_3 resin uptake test (RT_3I), which is merely a test for assessing the capacity for the serum binding proteins to bind thyroid hormones (T_3 or T_4), is not by itself a measure of the T_3 level in serum.

Measurement of Reverse T_3

Reverse T_3 (rT_3) is measured in whole serum by radioimmunoassay. The normal range in adults is approximately 25-75 ng/mL of serum. This is approximately one-third the concentration of the total T_3 concentration. The test is not generally available and is of limited clinical usefulness. Almost all clinical situations can be evaluated without an rT_3 determination on the basis of the serum T_4 and T_3 and their respective free hormone indices. Of potential diagnostic value, however, is the observation that the rT_3 level is elevated in systemic illness when serum T_3 is depressed. In hypothyroidism, on the other hand, both T_3 and rT_3 levels are low.

Measurement of Thyroglobulin

Thyroglobulin is measured in serum by radioimmunoassay. This measurement is invalidated by the presence in serum of endogenous antithyroglobulin antibodies. The serum thyroglobulin level is normally very low; it is elevated in conditions of thyroid overactivity, in which its secretion is increased along with thyroid hormones. Excess thyroglobulin is also released into the serum when the follicular structure of the thyroid gland is damaged, as in subacute thyroiditis. In these conditions, however, determination of the serum thyroglobulin level is not of diagnostic- value. It is of much greater clinical importance in following patients who have had surgery for thyroid cancer. Following total thyroidectomy, thyroglobulin should be undetectable in serum. The de novo appearance of this compound is good evidence that thyroid metastases have developed. In patients with existing thyroid metastases, an increase in the concentration of serum thyroglobulin usually indicates progression of the disease and vice versa.

Measurement of Protein-Bound Iodine (PBI)

Until 2 decades ago, PBI was a major test for measurement of serum thyroid hormone levels. It measures all precipitable iodine-containing proteins. Normally, the iodine content of serum T_4 approximates the PBI. Thus, about 66% of the weight of T_4 is iodine The normal PBI range is 4-8 μg/dL of serum, about two-thirds of the normal range for the total T_4 concentration. The PBI has been superseded by more specific- tests for thyroid hormones, but very occasionally it is still of value. For example, in congenital defects of thyroid hormonogenesis, the PBI may be considerably higher than the serum total T_4 iodine concentration. This indicates the secretion of nonhormonal organic iodine compounds. A similar phenomenon may occur in Hashimoto's thyroiditis. In addition, a variety of organic iodine x-ray contrast materials that may be interfering with thyroid function can be detected by this procedure.

Tests of Thyroid Gland Function

Radioactive Iodine Uptake (RAIU)

The rationale for this test was described in the section on iodide metabolism. Sodium iodide I 131—or, more recently, sodium iodide I 123—is administered orally, and a gamma scintillation counter is used to measure radioactivity over the area of the thyroid after 24 hours. This interval is chosen because the tracer uptake is at 24 hours near maximum. An earlier measurement is usually done as well, either 4 or 6 hours after radioisotope administration, because the RAIU may occasionally reach a maximum before 24 hours and subsequently decline by 24 hours. Situations in which this decline may be seen include severe thyrotoxicosis with rapid thyroid iodine turnover and early primary thyroid failure (such as in Hashimoto's thyroiditis), where a small remnant of functional thyroid under TSH stimulation is releasing newly synthesized hormone very rapidly without a long period of storage.

Uncommonly, hyperthyroidism may be associated with a very low or absent thyroid RAIU, which is of diagnostic importance in establishing the condition responsible for the hyperthyroidism. Examples of low RAIU hyperthyroidism are (1) subacute thyroiditis, (2) "spontaneously resolving hyperthyroiditis," (3) thyrotoxicosis factitia, (4) struma ovarii, (5) excessive iodide intake (jodbasedow effect), and (6) ectopic functional thyroid metastases after total thyroidectomy.

Perchlorate Discharge

The efficiency of the thyroid organification mechanism may be tested by giving $KClO_4$, 0.5 g in solution orally. 2-3 hours after oral administration of radioiodide, and observing its effect on the RAIU. Perchlorate inhibits the iodide transport mechanism. In subjects with a normal organification mechanism, ClO_4^- blocks further uptake of radioiodide by the thyroid but does not discharge more than 5% of the accumulated iodide in the next hour. Conversely, if iodide organification is incomplete and inorganic iodide is "backed up" within the thyroid cell, ClO_4^- blocks further I^- transport, but diffusion of I^- out of the cell continues. This is seen as discharge of radioactive iodide from the thyroid gland, as measured by detectable radioactivity over the gland. A positive test may be observed in (1) congenital iodide organification defects, (2) some cases of Hashimoto's thyroiditis, (3)

Graves' disease previously treated with radioactive iodide, and (4) patients who have ingested inhibitors of iodide organification such as methimazole and propylthiouracil. Sensitivity of the test may be enhanced by the administration of 1 mg of stable (nonradioactive) iodide together with the radioactive iodide. This provides additional substrate to stress the organification mechanism. The ClO_4^- discharge test is rarely used, because of its limited diagnostic value given the wide variety of more specific and simpler tests available.

Thyroid Imaging

The size and shape and some features of the internal structure of the thyroid may be assessed by a wide variety of thyroid imaging techniques. Their clinical application will be discussed in greater detail in subsequent sections of this chapter. However, they may be summarized as follows:

Radionuclide scan

After the administration of radioactive iodide—eg, sodium iodide I 123—an image of the thyroid may be obtained using an appropriate scanning apparatus. With the rectilinear scanner, the scanner moves in a matrix over the patient's neck, and the radioactivity at each point is recorded as a dot on paper or x-ray film. The advantage of this scan is that it accurately represents the size of the thyroid arid accurately localizes palpable thyroid abnormalities. On the other hand, its resolution is relatively poor, and it takes considerable time for the scan to be completed.

Over recent years, the pinhole collimated gamma camera (scintillation camera) has moved to the fore. The field of the gamma camera is relatively large and encompasses the entire thyroid. This allows a rapid scan of the thyroid without the necessity for moving the camera. The data are usually recorded on Polaroid film. Besides speed, the major advantage of this method is increased resolution.

Technetium Tc 99m pertechnetate is also concentrated by the thyroid and can be used for imaging with high resolution. Unlike radioactive iodine, this isotope is not organified and stored, and the scan is therefore performed very shortly (eg, 20 minutes) after isotope administration. Advantages of this scan in addition to its speed include the fact that it is not interfered with by drugs that block iodide organification. Occasionally, however, thyroid nodules that are "cold" with sodium iodide I 123 are not identified with technetium Tc 99m pertechnetate.

Fluorescent scan

The fluorescent scan provides an image of the distribution of nonradioactive, stable iodine within the thyroid. An approximate quantitation can also be made of the thyroidal iodine store. No radioisotope is administered. An external source of ^{241}Am is directed at the thyroid, resulting in the emission of x-rays by thyroid molecules. By selectively recording emitted 28.5 keV x-rays specific for iodine, an image of the intrathyroidal iodine distribution can be formed. The scan is similar in appearance to the rectilinear radioactive iodine scan. The advantage of the fluorescent scan is that it does not require radioactive iodine administration, and the thyroid may be imaged even if the RAIU is zero, such as when the patient is taking exogenous thyroid hormones. Thus, it can be used to determine whether thyroid tissue is absent or when it is present but not trapping radioiodine.

Ultrasonography

The application of this technique is similar to that for other organs. Its major use is to differentiate between cystic and solid thyroid lesions that may both appear to be "cold" on radionuclide scanning.

Tests of the Effects of Thyroid Hormone on Peripheral Tissues

The ultimate test of whether a patient is experiencing the effects of too much or too little thyroid hormone is not a measurement of hormone concentration in the blood or the size and functional activity of the thyroid but the effect of thyroid hormones on the peripheral tissues. Unfortunately, no simple, reproducible, and specific tests are available for this purpose.

Basal Metabolic Rate (BMR)

This test measures a resting subject's oxygen consumption with a spirometer. Oxygen consumption is increased in hyperthyroidism and vice versa. Because of the wide variability of results, attempts were made to standardize the test by correcting for body surface area, age, and sex and by making sure patients were truly at rest. The test is expensive, time-consuming, subject to a variety of influencing factors, and not very sensitive and is rarely used today.

Photomotogram

This simple apparatus can be used to measure the speed of relaxation of the Achilles tendon reflex. The patient kneels on a chair with the foot positioned in such a way as to interrupt a photoelectric light beam. The Achilles tendon is tapped with a hammer, which initiates a timer, and when the foot returns to its original position the timer is stopped and indicates the length of time taken for return of the foot to the original position. The approximate half-relaxation time is 230-350 ms. This test is of limited value in hypothyroidism, in which the relaxation time is typically prolonged. Although relaxation time is faster in hyperthyroidism, there is a large overlap with the normal range and the test is therefore of little clinical value. The test is influenced by a wide variety of other neuromuscular conditions as well as by pharmacologic agents, and in the author's opinion it is of limited diagnostic value.

Q-Kd Interval

Measurement of the Q-Kd interval provides an assessment of the effect of thyroid hormones on the rate of myocardial contraction. The test measures time elapsed between initiation of the QRS complex on the ECG and the arrival of the pulse wave in the brachial artery as measured by the Korotkoff sound. The Q-Kd interval is typically decreased in hyperthyroidism and increased in hypothyroidism. The limitations of the test are similar to those described for the photomotogram, and it too is rarely used.

Biochemical Changes

Thyroid hormones influence the concentration of a vast array of tissue and blood compounds as well as the activity of many different enzymes. Unfortunately, however, none of these changes are specific enough, sensitive enough, or convenient enough to measure to make them of clinical importance. Perhaps the most useful is the serum cholesterol, which is elevated in hypothyroidism and decreased in hyperthyroidism. Serum creatine phosphokinase (CPK) and lactate dehydrogenase (LDH) may also be increased in hypothyroidism.

Evaluation of the Hypothalamic-Pituitary-Thyroid Axis

Measurement of Serum TSH

Factors that influence TSH secretion have been described above. TSH is measured in blood by radioimmunoassay. This test is the most sensitive, convenient, and specific test for the diagnosis of primary hypothyroidism. The TSH assay is, however, not sensitive enough to detect TSH in the serum of many euthyroid patients and therefore cannot distinguish between euthyroidism and hyperthyroidism. There is also increasing evidence that the measurement of TSH by radioimmunoassay does not always correlate precisely with the degree

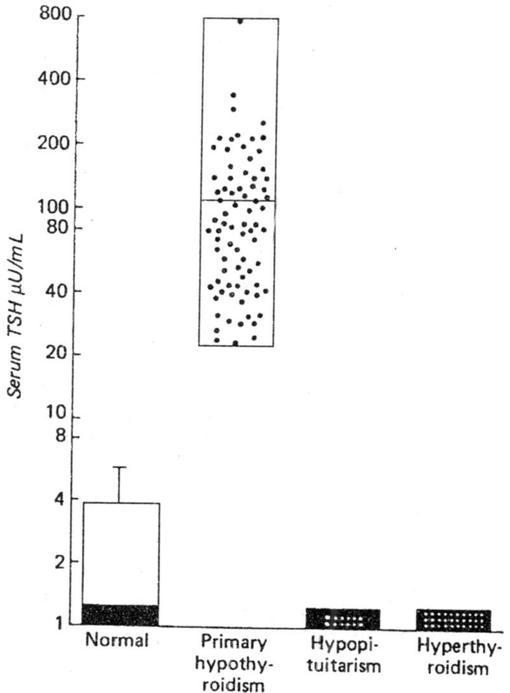

Fig. 7.23. Diagnostic value of serum TSH concentration in primary hypothyroidism.

of TSH bioactivity. Clinically, however, this is not important, and there is generally an excellent correlation between TSH immunoactivity and biologic activity.

Thyrotropin-Releasing Hormone (TRH) Test

In a sense, the TRH test may be regarded as a test of the metabolic effects of thyroid hormones on the peripheral tissues—in this case, specifically the thyrotrophs. In hyperthyroidism, the TSH response to TRH is dampened or abolished, and in primary hypothyroidism the maximum response is enhanced. The TSH response to TRH is slightly greater in females than in males. A normal response is a 2-to 5-fold increase over basal serum TSH 15-30 minutes after intravenous injection of 500 μg of TRH.

TRH is usually administered intravenously (400 or 500 μg as a single bolus), but it may also be given orally, intramuscularly, or by continuous intravenous infusion. When TRH is given intravenously as a bolus injection, the serum TSH level increases rapidly, reaching a maximum 15-30 minutes after injection, and then declines more slowly over a period of a few hours. The greatest value of this test is in the diagnosis of mild hyperthyroidism, for which it is the most sensitive test. As an indication of its sensitivity, the administration of exogenous thyroid hormones to normal subjects—at a dose insufficient to raise serum hormone levels above the normal range—suppresses the TSH response to TRH stimulations. The test is of less value in primary hypothyroidism (thyroidal failure), in which the high, unstimulated TSH value is in itself sufficient evidence for the diagnosis. While the TSH response to TRH is augmented in hypothyroidism, this increase is generally proportionate to the elevation in the basal TSH concentration and therefore provides little additional information.

The TRH test is the only test available to distinguish between secondary (pituitary) and tertiary (hypothalamic) hypothyroidism. In the former, there is a decreased or absent TSH response to TRH stimulation. In the latter, despite a low serum TSH in association with low serum T_4 and T_3 levels, there is an adequate—but sometimes delayed—TSH response to TRH injection.

TSH Stimulation Test

This test measures the responsiveness of the thyroid to exogenous TSH. Originally used to test the degree of thyroid reserve in primary hypothyroidism, it has been superseded by direct measurement of the endogenous TSH level in serum. The test is time-consuming, expensive, and potentially dangerous. Bovine TSH (10 units) is administered intramuscularly daily for 3 days. The thyroid RAIU is measured before and after TSH administration. It is important to skin test patients before administration of bovine TSH, because serious allergic responses have occurred.

Thyroid Suppression With T_3

In Graves' disease, thyroid function is autonomous and no longer under the control of the hypothalamic-pituitary axis. Because thyroid, overactivity is produced by a factor other than TSH, suppression of endogenous TSH by administration of T_3 will not reduce the thyroid RAIU. The standard T_3 suppression test involves the initial measurement of the thyroid RAIU followed by the administration for 5-7 days of 100 μg of T_3 per day in divided doses. After this period, the RAIU is measured again. T_3 suppression should reduce the RAIU in normal subjects by at least 50%. Although the T_3 suppression test and the TRH test are both useful in the evaluation of selected hyperthyroid patients, there are important differences between them. The TSH response to TRH is blunted when the pituitary senses excess thyroid hormone, irrespective of the condition producing the hyperthyroidism. In contrast, the T_3 suppression test is not used to diagnose thyrotoxicosis but rather thyroid autonomy. Nonsuppressibility of the RAIU is not sufficient for the diagnosis of hyperthyroidism. For example, after radioactive iodine treatment for Graves' disease, the patient may be euthyroid bill still have a T_3-nonsuppressible thyroid.

Measurement of Pathologic Thyroid Antibodies

Antithyroglobulin and Antimicrosomal Antibodies

These antibodies are present in the serum of patients with autoimmune thyroid disease and to a lesser degree in some other thyroid conditions. The most commonly used method for measuring these antibodies is by a hemagglutination technique. Thyroglobulin or thyroid microsomal proteins are absorbed onto tannic acid-treated sheep red blood cells. Antibodies against these antigens agglutinate the cells, and dilution of the patient's serum allows determination of antibody titer. Thyroglobulin antibodies and antimicrosomal antibodies may also be measured by radioimmunoassay, which is more sensitive.

Thyroid-Stimulating Immunoglobulins (TSI)

Many different tests have been used for the measurement of TSI. Most have been too impractical or insensitive for general clinical use.

Long-acting thyroid stimulator (LATS)

This was the initial assay for the measurement of TSI. The assay measures the ability of patients' serum to release radioiodine from the thyroid glands of mice prelabeled with radioiodine. The difficulties with this assay are that it is time-consuming, expensive, and of limited sensitivity and specificity. The false-negative rate is approximately 50%. Many other factors also influence the bioassay animals, leading to difficulty in interpreting the results.

LATS protector

This is an adaptation of the LATS assay. Advantage is taken of the species-specificity of TSI. Thus, human thyroid tissue absorbs LATS activity from LATS-positive serum. Serum or IgG from a patient with active Graves' disease prevents adsorption of LATS to the thyroid tissue and therefore "protects" it. Different human sera can therefore be assessed in terms of their efficacy in protecting a standard LATS serum of known activity from adsorption to thyroid tissue. The LATS is measured by the mouse bioassay method described above. The disadvantage of the LATS protector assay is that it is even more laborious than the LATS assay. However, false-negatives in Graves' disease are reduced, and the test would be of clinical value if it were easier to perform.

Thyroid cAMP assays

Serum— or IgG—from patients with Graves' disease is tested for its ability to stimulate cAMP generation in human thyroid tissue—either slices or plasma membranes. Initial studies have indicated a 10-30% false-negative rate. A major disadvantage of the slice assay is the need for fresh human thyroid tissue. Variations on the assay include quantitation of colloid droplets in thyroid follicular cells and measurement of thyroid hormones secreted by the tissue slices. None of these assays are practical enough for extensive clinical use. Recently, the cAMP response in human thyroid cells cultured in monolayers has been adapted for the bioassay of TSI. The sensitivity and practicality of this new assay will now permit the routine determination of TSI.

TSH binding inhibition (TBI)

This assay utilizes the binding of radiolabeled bovine TSH to human thyroid plasma membrane preparations. This binding is inhibited to varying degrees by serum or IgG from patients with active Graves' disease. It is important to realize that the TBI assay is not truly a TSI assay in that it does not measure biologic activity. Another name for the test is TSH displacement assay (TDA). While this assay is convenient and has a relatively low false-negative rate, it is plagued by false-positive results. Thus, a wide variety of nonspecific interfering substances, such as thyroglobulin, inhibit TSH binding. There is also a low correlation between the results obtained in the TBI assay and in the thyroid slice cAMP assay using the same, test sera. Clinical use of the TBI assay is therefore declining.

Disorders of the Thyroid Gland

Patients with thyroid disease will usually complain of (1) thyroid enlargement, or goiter; (2) symptoms of thyroid deficiency, or hypothyroidism: (3) symptoms of thyroid hormone excess, or hyperthyroidism; or (4) complications of a specific form of hyperthyroidism, Graves' disease, which may present with striking prominence of the eyes (exophthalmos) or, rarely, thickening of the skin over the lower legs (thyroid dermopathy).

The *history* should include evaluation of symptoms related to the above complaints, discussed in more detail below. Exposure to ionizing radiation in childhood has been associated with an increased incidence of thyroid disease, including cancer. Iodide ingestion in the form of kelp or iodide-containing cough preparations may induce goiter, hypothyroidism, or hyperthyroidism. Lithium carbonate, used in the treatment of manic-depressive psychiatric disorder, can also induce hypothyroidism and goiter. Residence in an area of low dietary iodide is associated with iodine-deficiency goiter, or "endemic goiter." Although in developed countries dietary iodide is generally adequate, there are still areas low in natural iodine, in undeveloped countries in Africa, Asia, South America, and inland mountainous areas. Finally, the family history should be explored with particular reference to goiter, hyperthyroidism, or hypothyroidism, as well as immunologic disorders such as diabetes, rheumatoid disease, pernicious anemia, or myasthenia gravis, which may be associated, with an increased incidence of thyroid disease. Multiple endocrine neoplasia type II (Sipple's syndrome) with medullary carcinoma of the thyroid gland is an autosomal dominant condition.

The thyroid is firmly attached to the anterior trachea midway between the sternal notch and the thyroid cartilage; it is often easy to see and to palpate. The patient should have a glass of water for comfortable swallowing. There are 3 maneuvers: (1) With a good light coming from behind the examiner, have the patient swallow a sip of water. Observe the gland as it moves up and down. Enlargement and nodularity can often be noted. (2) Palpate the gland anteriorly. Gently press down with one thumb on one side of the gland to rotate the other lobe forward, and palpate as the patient swallows. (3) Palpate the gland from behind the patient with the middle 3 fingers on each lobe while the patient swallows. An outline of the gland can be traced on the skin of the neck and measured. Nodules can be measured in a similar way. Thus, changes in the size of the gland or in nodules can be followed.

Each lobe of the normal thyroid gland measures about 2 cm in vertical dimension and about 1 cm in horizontal dimension above the isthmus. An enlarged thyroid gland has been called *goiter*. Generalized enlargement is termed *diffuse goiter*; irregular or lumpy enlargement is called *nodular goiter*.

Hypothyroidism

Hypothyroidism is a clinical syndrome resulting from a deficiency of thyroid hormones, which in turn results in a generalized slowing down of metabolic processes. Hypothyroidism in infants and children results in marked slowing of growth and development, with serious permanent consequences including mental retardation. Hypothyroidism with onset in adulthood causes a generalized slowing down of the organism, with the deposition of glycosaminoglycans in intracellular spaces, particularly in skin and muscle, producing the clinical picture of *myxedema*. The symptoms of hypothyroidism in adults are largely reversible with therapy.

Etiology

Hypothyroidism may be classified in a number of ways. It may be primary (thyroid failure), secondary (to pituitary TSH deficit), or tertiary (due to hypothalamic deficiency of TRH); or there may be an abnormality of the thyroxine (T_4) receptor in the cell, inducing peripheral resistance to the action of thyroid hormones. Hypothyroidism can also be classified as goitrous or nongoitrous, but this is probably unsatisfactory since Hashimoto's thyroiditis may produce hypothyroidism with or without goiter. The incidence of various causes of hypothyroidism will vary depending on geographic and environmental factors such as dietary iodide and goitrogen intake, the genetic characteristics of the population, and the age distribution of the population (pediatric or adult).

Hashimoto's thyroiditis is probably the most common cause of hypothyroidism. In younger patients, it is more likely to be associated with goiter; in older patients, the gland may be totally destroyed by the immunologic process, and the only trace of the disease will be a persistently positive test for antithyroid microsomal antibodies. Similarly, the end stage of Graves' disease is hypothyroidism. This is accelerated by destructive therapy such as administration of radioactive iodine or subtotal thyroidectomy. Thyroid glands involved in autoimmune disease are particularly susceptible to excessive iodide intake, such as ingestion of kelp tablets, iodide-containing cough preparations, or administration of iodide-containing radiographic contrast media. In these situations, the large amounts of iodide block thyroid hormone synthesis, producing hypothyroidism with goiter. Although the process may be temporarily reversed by withdrawal of iodide, the underlying disease will often progress, and permanent hypothyroidism will usually supervene. Hypothyroidism may occur during the late phase of subacute thyroiditis; this is usually transient, but it is permanent in about 10% of patients. Iodide deficiency is rarely a cause of hypothyroidism in the USA but may be more common in underdeveloped countries. Certain drugs can block hormone synthesis and produce hypothyroidism with goiter; at present, the most common pharmacologic cause of hypothyroidism (other than iodide) is lithium carbonate, used for the treatment of manic-depressive states. The antithyroid drugs propylthiouracil and methimazole in continuous dosage will do the same. Inborn errors of thyroid hormone synthesis will result in severe hypothyroidism if the block in hormone synthesis is complete, or mild hypothyroidism if the block is partial.

Pituitary and hypothalamic deficiencies as causes for hypothyroidism are quite rare and are usually associated with other symptoms and signs. Peripheral resistance to thyroid hormones is extremely rare, occurs in families, and is characterized by high serum concentrations of T_4 and T_3, goiter, and features of hypothyroidism rather than hyperthyroidism.

Pathogenesis

Thyroid hormone deficiency affects every tissue in the body, so that the symptoms are multiple. Pathologically, the most characteristic finding is the accumulation of glycosaminoglycans—mostly hyaluronic acid—in interstitial tissues. Accumulation of this hydrophilic substance accounts for the interstitial edema that is particularly evident in the skin, heart muscle, and striated muscle. The accumulation is due not to excessive synthesis but to decreased destruction of glycosaminoglycans.

Clinical Presentations

Newborn infants (Cretinism)

The term cretinism was originally applied in endemic goiter areas to infants with mental retardation, short stature, a characteristic puffy appearance of the face and hands, and frequently deaf mutism and pyramidal tract signs. In the USA, neonatal hypothyroidism may occur following administration during pregnancy of iodides, antithyroid drugs in large doses, or radioactive iodine for thyrotoxicosis. In addition, about one out of every 7000 live births is associated with athyreosis, ie, no thyroid gland develops. Some patients considered athyreotic have been shown to have ectopic thyroid tissue in the developmental tract between the base of the tongue and the anterior portion of the neck, so that cretinism in these infants is due not to lack of development of the thyroid but rather due to impaired thyroid development associated with thyroid ectopia.

The symptoms of hypothyroidism in newborns include respiratory difficulty, cyanosis, jaundice, poor feeding, hoarse cry, umbilical hernia, and marked retardation of bone maturation. The proximal tibial epiphysis and distal femoral epiphysis are present in almost all full-term infants with a body weight of over 2500 g. Absence of these epiphyses strongly suggests hypothyroidism. Because early treatment is essential to prevent permanent mental retardation, neonatal screening programs have been introduced to measure serum T_4 or TSH: A serum T_4 under 6 μg/dL, or a serum TSH over 30 μU/mL is indicative of neonatal hypothyroidism. The diagnosis can then be confirmed by radiologic evidence of retarded bone age.

Children

Hypothyroidism in children is characterized by retarded growth and evidence of mental retardation. In the adolescent, precocious puberty may occur, and there may be enlargement of the sella turcica in addition to short stature. This is not due to pituitary tumor but probably to pituitary hypertrophy associated with excessive TSH production.

Adults

In adults, the common features of hypothyroidism include easy fatigability, coldness, weight gain, constipation, menstrual irregularities, and muscle cramps. Physical findings include a cool, dry skin, puffy face and hands, hoarse, husky voice, and slow reflexes. Reduced conversion of carotene to vitamin A and increased blood levels of carotene may give the skin a yellowish colour.

Cardiovascular signs include bradycardia and cardiac enlargement; the enlargement may be due in part to interstitial edema and left ventricular dilatation but is often due to pericardial effusion. The ECG reveals low voltage of QRS complexes and P and T waves, with improvement on therapy. The degree of pericardial effusion can easily be determined by echocardiography. Cardiac output is reduced, although congestive heart failure and pulmonary edema are rarely noted. Although there is controversy about whether myxedema induces coronary artery disease, there is evidence that coronary artery disease is more common in patients with hypothyroidism, particularly in older patients. For this reason, it is important to start levothyroxine replacement therapy in myxedematous patients over age 50 with very small doses, increasing the dose slowly over a period of several months.

Pulmonary function in adult hypothyroidism is characterized by shallow, slow respirations and impaired ventilatory responses to hypercapnia or hypoxia. This feature is important in the development of myxedema coma.

Intestinal peristalsis is markedly slowed, resulting in chronic constipation and occasionally severe fecal impaction.

Renal function is impaired, with decreased glomerular filtration rate and impaired ability to excrete a water load. This predisposes the myxedematous patient to water intoxication if excessive free water is administered.

There are at least 4 mechanisms that may contribute to *anemia* in patients with hypothyroidism: (1) impaired hemoglobin synthesis as a result of thyroxine deficiency; (2) iron deficiency from increased iron loss with menorrhagia, as well as impaired intestinal absorption of iron: (3) folate deficiency from impaired intestinal absorption of folic acid: and (4) pernicious anemia, with vitamin B_{12}-deficient megaloblastic anemia. The pernicious anemia is often part of a spectrum of autoimmune diseases: myxedema due to chronic thyroiditis, with antithyroid antibodies present; pernicious anemia with anti-parietal cell antibodies present; diabetes mellitus with anti-islet cell antibodies present; adrenal insufficiency with anti-adrenal antibodies present; etc.

Many patients complain of symptoms referable to the neuromuscular system, eg, severe muscle cramps, paresthesias, and muscle weakness.

Central nervous system symptoms include chronic fatigue, lethargy, and inability to concentrate.

Hypothyroidism produces impaired peripheral metabolism of estrogens, with altered FSH and LH secretion, resulting in anovulatory cycles and infertility. This may also be associated with severe menorrhagia.

Patients with myxedema are usually quite placid but can be severely depressed or even extremely agitated ("myxedema madness").

Diagnosis

The laboratory diagnosis of hypothyroidism is not difficult. A low serum T_4, low resin T_3 uptake, and elevated serum TSH are characteristic of primary hypothyroidism. Serum T_3 levels are variable and may be within the normal range. A positive test for antithyroid antibodies suggests underlying Hashimoto's thyroiditis. In patients with pituitary myxedema, the free thyroxine index will be low but serum TSH will not be elevated. To differentiate pituitary from hypothalamic disease, the TRH test is most helpful. An elevated basal TSH level with excessive response indicates primary hypothyroidism. Absence of TSH response to TRH indicates pituitary deficiency. A partial or "normal" response indicates that pituitary function is intact but that a defect exists in hypothalamic secretion of TRH. Since, pituitary or hypothalamic hypothyroidism is rare, TRH tests are usually unnecessary in the diagnosis of hypothyroidism.

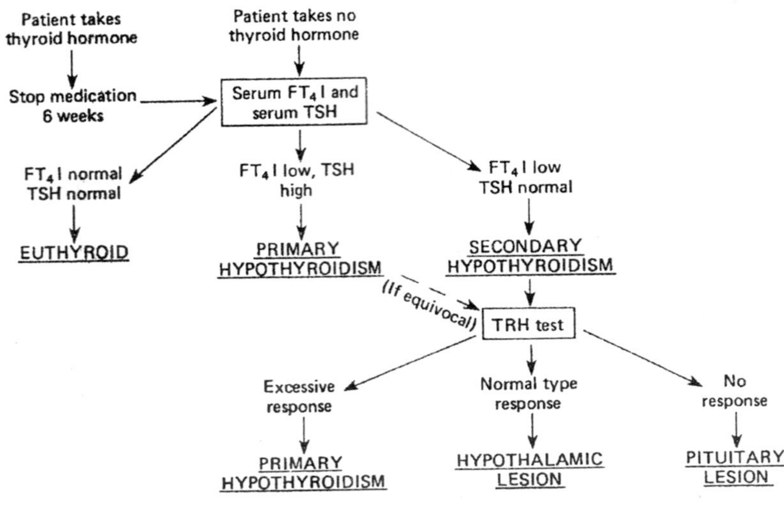

Fig. 7.24. Diagnosis of hypothyroidism.

The patient may be taking thyroid medication (levothyroxine or desiccated thyroid tablets) when first seen. A positive test for antithyroid antibodies would suggest underlying Hashimoto's thyroiditis, in which case the medication should be continued. If antibodies are absent, the medication should be withdrawn for 6 weeks and FT_4I and TSH determined. The 6-week period of withdrawal is necessary because of the long half-life of thyroxine (8 days) and to allow the pituitary gland to recover after a long period of suppression. In hypothyroid individuals, TSH becomes markedly elevated at 5-6 weeks and T_4 remains subnormal, whereas both are normal after 6 weeks in euthyroid controls.

In patients with hypothyroidism, a delay in muscle contraction and relaxation is usually evident on physical examination, particularly in the biceps reflex. The speed of contraction and partial relaxation of the Achilles tendon reflex can be measured by the "photomotogram". The response in hypothyroidism (300-600 ms) overlaps the normal response (230-350 ms), so that the test is not useful for the diagnosis of hypothyroidism. However, the test is quite useful to follow patients under therapy. Improvement in reflex contraction and relaxation time will coincide with the clinical response to appropriate therapy.

The clinical picture of fully developed myxedema is usually quite clear, but the symptoms and signs of mild hypothyroidism may be very subtle. Patients with hypothyroidism will at times present with unusual features—neurasthenia with symptoms of muscle cramps, paresthesias, and weakness; anemia: disturbances in reproductive

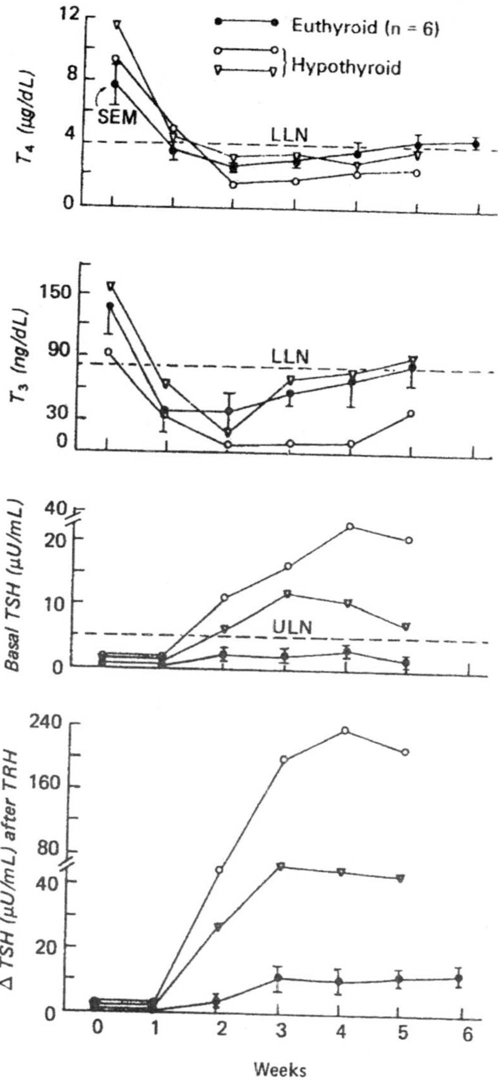

Fig. 7.25. Changes in T_4, T_3, TSH and TRH response following abrupt withdrawal of suppressive thyroxine therapy.

function, including infertility, delayed puberty, or menorrhagia; idiopathic edema or pleuropericardial effusions; retarded growth; obstipation: chronic rhinitis or hoarseness due to edema of nasal mucosa on vocal cords; and severe depression progressing to emotional instability or even frank paranoid psychosis. In such cases, the diagnostic studies outlined above will confirm or rule out hypothyroidism as a contributing factor.

Complications

Myxedema coma

Myxedema coma is the end stage of untreated hypothyroidism. It is characterized by progressive weakness, stupor, hypothermia, hypoventilation, hypoglycemia, hyponatremia, water intoxication, shock, and death. Although rare, it may occur more frequently in the future associated with the increasing use of radioiodine for the treatment of Graves' disease, with resulting permanent hypothyroidism. Since it occurs most frequently in older patients with underlying pulmonary and vascular disease, the mortality rate is extremely high.

The patient (or a family member if the patient is comatose) may recall previous thyroid disease, radioiodine therapy, or thyroidectomy. The medical history is of gradual onset of lethargy progressing to stupor or coma. Examination shows bradycardia and marked hypothermia, with body temperature as low as 24°C (75°F). The patient is usually an obese elderly woman with yellowish skin, a hoarse voice, a large tongue, thin hair, puffy eyes, ileus, and slow reflexes. There may be signs of other illnesses such as pneumonia, myocardial infarction, cerebral thrombosis, or gastrointestinal bleeding.

Laboratory clues to the diagnosis of myxedema coma include lactescent serum, high serum carotene, elevated serum cholesterol, and increased cerebrospinal fluid protein. Pleural, pericardial, or abdominal effusions with high protein content may be present. Serum tests will reveal a low T_4(RIA), low RT_3U, and usually a markedly elevated TSH. Thyroidal radioactive iodine uptake is low, and antithyroid antibodies are usually strongly positive, indicating underlying thyroiditis. The ECG shows sinus bradycardia and low voltage. Frequently if laboratory studies are not readily available, the diagnosis must be made clinically.

The pathophysiology of myxedema coma involves 3 major aspects: (1) CO_2 retention and hypoxia, (2) fluid and electrolyte imbalance, and (3) hypothermia. CO_2 retention has long been recognized as an integral part of myxedema coma and had been attributed to such factors as obesity, heart failure, ileus, immobilization, pneumonia, pleural or peritoneal effusions, central nervous system depression, and weak chest muscles. More recently, Zwillich et al have demonstrated that the ventilatory responses to hypoxia are markedly depressed in patients with myxedema, as are the ventilatory responses to hypercapnia. These authors suggest that the failure of the myxedema patient to respond to hypoxia or hypercapnia may be associated with hypothermia. Thyroid hormone therapy in patients with myxedema markedly improves the ventilatory response to hypoxia. Because of the impaired ventilatory drive, assisted respiration is almost always necessary in patients with myxedema coma. The major fluid and electrolyte disturbance is water intoxication, which presents as hyponatremia and is managed by water restriction. Hypothermia is frequently not recognized, because the ordinary clinical thermometer only goes down to about 34°C (94°F); a laboratory type thermometer that registers a broader scale must be used to obtain accurate body temperature readings. The low body temperature may be due to diminished thyroxine stimulation of the sodium-potassium transport mechanism and decreased ATPase activity. Active rewarming of the body is contraindicated, because it may induce vasodilatation and vascular collapse. A rise in body temperature is a useful indication of therapeutic effectiveness of thyroxine.

Other disorders that may precipitate myxedema coma include heart failure, pulmonary edema, pleural or peritoneal effusions, ileus, and anemia. Adrenal insufficiency occurs occasionally in association with myxedema coma, but it is relatively rare and usually associated with either pituitary myxedema or concurrent autoimmune adrenal insufficiency (Schmidt's syndrome). Drug intoxication is quite common, and myxedema coma that develops in the hospital may be precipitated by administration of sedatives or narcotics. Seizures, bleeding episodes, hypocalcemia, or hypercalcemia may be present.

It is important to differentiate pituitary myxedema from primary myxedema. In pituitary myxedema, adrenal insufficiency may be present, and adrenal replacement is essential. Clinical clues to the presence of pituitary myxedema include a history of amenorrhea or impotence and physical evidence of scanty pubic or axillary hair. Laboratory findings will usually reveal normal serum cholesterol and normal or low pituitary TSH levels. On skull x-ray, the sella turcica may be enlarged. The treatment of myxedema coma is discussed below.

Myxedema heart disease

Hypothyroidism has a direct effect upon the myocardium, causing, interstitial edema, nonspecific myofibrillary swelling, and pericardial effusion, manifested by bradycardia and diminished cardiac output. These complications are usually completely reversible with appropriate therapy.

Myxedema and coronary artery disease

Since myxedema frequently occurs in older persons, it is frequently associated with underlying coronary artery disease. In this situation, the low levels of circulating thyroid hormone actually protect the heart against increased demands that would result in increasing angina pectoris or myocardial infarction. Correction of the myxedema must be done very cautiously in order not to provoke arrhythmia, angina, or acute myocardial infarction. In some patients, it is necessary to perform aortic coronary bypass surgery to improve coronary circulation before full replacement therapy can be instituted.

Treatment of Hypothyroidism

Hypothyroidism is treated with levothyroxine, which is available in pure form and stable and inexpensive. Levothyroxine is converted in the body in part to T_3, so that both hormones become available even though only one is administered. Desiccated thyroid is unsatisfactory because of its variable hormone content, and triiodothyronine (as liothyronine) is unsatisfactory because of its rapid absorption and rapid disappearance from the bloodstream. The half-life of levothyroxine is about 8 days, so it need be given only once daily. Although only 40-60% of the preparation is absorbed, blood levels are easily monitored by following the free thyroxine index and serum TSH levels. The average replacement dose of levothyroxine in adults is 0.1-0.3 mg/d, with a mean of 0.18 mg/d. In infants under 1 year of age, the dose of levothyroxine is 6 μg/kg/d; in older children, the dosage is 3-4 μg/kg/d.

Treatment of Myxedema With Heart Disease

In long-standing hypothyroidism or in older patients, particularly those with cardiovascular disease, it is imperative to start treatment slowly. Levothyroxine is given in a dosage of 0.025 mg/d for 2 weeks, increasing by 0.025 mg every 2 weeks until a dose of 0.1 or 0.15 mg daily is reached. It usually takes about 2 months for a patient to come into equilibrium on full dosage. In these patients, the heart is very sensitive to the level of circulating thyroxine, and if angina pectoris or cardiac arrhythmia develops, it is essential to reduce the dose of thyroxine immediately. In younger patients, or patients with mild disease, full replacement may be started immediately.

Treatment of Myxedema Coma

Myxedema coma is an acute medical emergency and should be treated in the intensive care unit. Blood gases must be monitored regularly, and the patient usually requires intubation and mechanical ventilation. Associated illnesses such as infections must be treated by appropriate therapy. Intravenous fluids should be administered with caution, and excessive free water intake must be avoided.

Because patients with myxedema coma absorb all drugs poorly, it is imperative to give levothyroxine intravenously. Holvey pointed out that these patients have marked depletion of serum thyroxine and a large number of empty binding sites on thyroxine-binding globulin and therefore should receive an initial loading dose of thyroxine intravenously followed by a small daily intravenous dose. An initial dose of 300-400 μg of levothyroxine is administered intravenously, followed by 50 μg of levothyroxine intravenously daily. This preparation is commercially available. Intravenous triiodothyronine (liothyronine) is not commercially available, and fresh, sterile solutions of liothyronine in alkaline buffer must be prepared daily. The clinical guides to improvement are a rise in body temperature and the return of normal cerebral function. Hydrocortisone hemisuccinate, 100 mg intravenously followed by 50 mg intravenously every 6 hours, is indicated if the patient has associated adrenal or pituitary insufficiency—and indeed has been recommended for all patients with myxedema coma. If the adrenal or pituitary status is not known, it is probably desirable to support the patient with intravenous hydrocortisone, but adrenal function is probably adequate in most patients with myxedema coma.

When giving levothyroxine intravenously in large doses, there is an inherent risk of precipitating angina, heart failure, or arrhythmias in older patients with underlying coronary artery disease. Thus, this type of therapy is not recommended for ambulatory patients with myxedema; it is better to start slowly and buildup the dose as noted above.

Course and Prognosis

The course of untreated myxedema is one of slow deterioration leading eventually to myxedema coma and death. The results of treatment are very gratifying. Because of the long half-life (8 days) of thyroxine, it takes time to establish equilibrium on a fixed dose. Therefore, it is important to monitor the free thyroxine index and TSH levels every 4-6 weeks until a normal balance is reached. Thereafter, the patient must be maintained on replacement therapy for life, with free thyroxine index and TSH monitoring about once a year.

The mortality rate of myxedema coma was about 80% at one time. The prognosis has been vastly improved by recognition of the importance of mechanically assisted respiration and the use of intravenous levothyroxine. At present, the outcome probably depends upon how well the underlying disease problems can be managed.

Hyperthyroidism and Thyrotoxicosis

Thyrotoxicosis is the clinical syndrome that results when tissues are exposed to high levels of circulating thyroid hormone. In most instances, thyrotoxicosis is due to hyperactivity of the thyroid gland, or hyperthyroidism. Occasionally, thyrotoxicosis may be due to other causes such as excessive ingestion of thyroid hormone or excessive secretion of thyroid hormone from ectopic sites.

Diffuse Toxic Goiter (Graves' Disease)

Graves' disease is the most common form of thyrotoxicosis and may occur at any age, more commonly in females than in males. The syndrome consists of one or more of the following features: (1) thyrotoxicosis, (2) goiter, (3) ophthalmopathy (exophthalmos), and (4) dermopathy (pretibial myxedema).

Etiology

The cause of Graves' disease is not known. There is a strong familial predisposition in that about 15% of patients with Graves' disease have a close relative with the same disorder, and about 50% of relatives of patients with Graves' disease have circulating antithyroid antibodies. Females are involved about 5 times more commonly than males. The disease may occur at any age, with a peak incidence in the 20- to 40-year age group. Human leukocyte antigen (HLA) typing has indicated an increased frequency of HLA-B8 in Occidental and HLA-Bw35 in Oriental patients with Graves' disease and in their relatives, indicating a genetic factor in the pathogenesis of the disease.

Pathogenesis

The present concept of the pathogenesis of Graves' disease is that it is an autoimmune disease in which T lymphocytes become sensitized to antigens within the thyroid gland and stimulate B lymphocytes to synthesize antibodies to these antigens. One such antibody may be directed against the TSH receptor site in the thyroid cell membrane and has the capacity to stimulate the thyroid cell to increased growth and function. This antibody has been called thyroid-stimulating immunoglobulin (TSI). Initially it was measured by bioassay in mice and was called long-acting thyroid stimulator (LATS). However, LATS could be detected in this heterologous assay in only about 50% of patients. When human thyroid tissue was used for detection of this

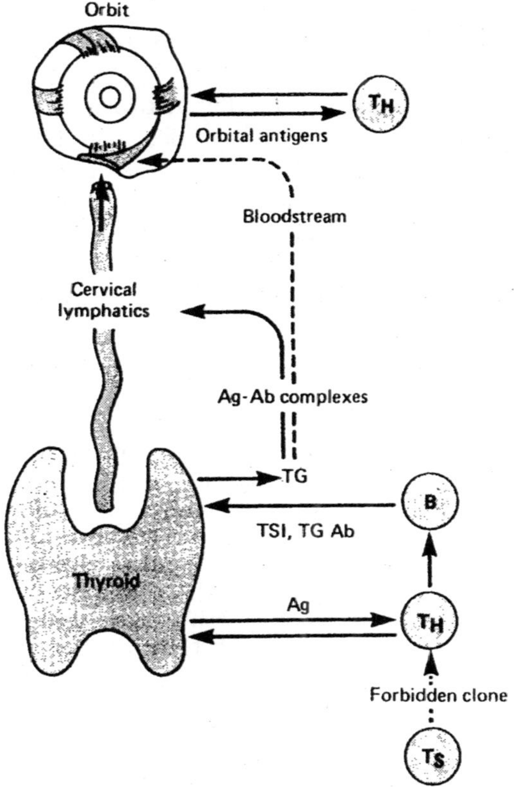

Fig. 7.26. Theory of the pathogenesis of Graves' disease.

antibody (either by increase in colloid droplet formation or increase in cAMP production), TSI could be detected in over 90% of patients with Graves' disease. The presence of this circulating antibody was positively correlated with active disease and with relapse of the disease. It is not clear what "triggers" the acute episode. Volpe has suggested that there is a defect in immunoregulation with failure of "suppressor" T lymphocyte function, allowing "helper" T lymphocytes to stimulate B lymphocytes to produce immunoglobulins directed against the thyroid gland, some of which would have the capacity to stimulate the thyroid gland as TSI. Kriss has proposed that the ophthalmopathy of Graves' disease is due to deposition in orbital muscles of immune complexes such as thyroglobulin-antithyroglobulin complexes, initiating an immune complex inflammatory reaction. An alternative theory for the pathogenesis of ophthalmopathy is that it is due to lymphocyte-mediated autoimmune disease involving orbital muscle. Both theories may be correct. The pathogenesis of the thyroid dermopathy (pretibial myxedema) and the rare subperiosteal inflammation has not yet been clarified.

Many of the symptoms of thyrotoxicosis suggest a possible synergism between epinephrine and thyroxine. Thus, there is tachycardia, tremor, sweating, lid lag, and stare. Yet circulating levels of epinephrine are normal. This situation has been clarified by the discovery of an increased number of epinephrine binding sites in heart muscle in animals treated with thyroxine, suggesting that the increased epinephrine sensitivity of patients with thyrotoxicosis may be due to these increased epinephrine receptors.

Clinical findings

Symptoms and signs

In younger individuals, common manifestations include palpitations, nervousness, easy fatigability, excessive sweating, preference for cold, hyperkinesia, and diarrhea. Thyroid enlargement, thyrotoxic eye signs, and mild tachycardia commonly occur. In addition, in children there is rapid growth with accelerated bone maturation. In older patients, cardiovascular and myopathic manifestations often predominate. Loss of muscle mass may be so severe that the patient cannot rise from a chair without assistance. In patients over the age of 60, the most common presenting complaints are palpitation, dyspnea on exertion, tremor, nervousness, and weight loss.

The eye signs of Graves' disease have been classified by the American Thyroid Association. Note that this classification utilizes the mnemonic "NO SPECS." Class 1 involves spasm of the upper lids associated with active thyrotoxicosis and usually resolves spontaneously when the thyrotoxicosis is adequately controlled. Classes 2 through 6 represent true infiltrative disease involving orbital muscles and orbital tissues. Class 2 represents soft tissue involvement with periorbital edema, congestion or redness of the conjunctiva, and swelling of the conjunctiva (chemosis). Class 3 represents proptosis as measured by the Hertel exophthalmometer. It consists of 2 prisms with a scale, mounted on a bar. The prisms are placed on the lateral orbital ridges, and the distance from the orbital ridge to the anterior cornea is measured on the scale. Class 4 represents muscle involvement. The muscle most commonly involved is the inferior rectus, which impairs upward gaze. The muscle second most commonly involved is the medial rectus with impairment of lateral gaze. Class 5 represents corneal involvement (keratitis), and class 6 loss of vision from optic nerve involvement. Infiltrative ophthalmopathy is due to infiltration of the extraocular muscles with lymphocytes and edema fluid in an acute inflammatory reaction. This produces muscle enlargement with proptosis, periorbital edema, chemosis, and congestion of the conjunctiva. In addition, the impaired muscle movement results in diplopia. Ocular muscle enlargement has been demonstrated beautifully in orbital CT scans. When muscle swelling occurs posteriorly, toward the apex of the orbital cone, the optic nerve is compressed, and it is when this happens that loss of vision is likely to occur.

Thyroid dermopathy consists of thickening of the skin, particularly over the lower tibia, due to accumulation of glycosaminoglycans. It is relatively rare, occurring in about 2-3% of patients with Graves' disease. It is usually associated with ophthalmopathy. The skin is markedly thickened and cannot be picked up between the fingers. Sometimes it involves the entire lower leg and may extend onto the feet. Bony involvement (osteopathy), with subperiosteal bone formation and swelling is particularly evident in the metacarpal bones. This too is a relatively rare finding. A more common finding in Graves' disease is separation of the nail from its bed, or onycholysis.

Laboratory findings

T_3 is secreted in excess quantity early in the development of Graves' disease and in recurrences of Graves' disease. Thus, high levels of T_3 by radioimmunoassay and an elevated thyroid radioactive iodine uptake are the characteristic laboratory findings in Graves' disease. In addition, resin T_3 uptake and serum T_4 levels will also be elevated. Antithyroid antibodies are usually present, particularly thyroid-stimulating immunoglobulin.

In patients with equivocal findings, it may be necessary to resort to special studies. The thyroid gland is normally suppressed when exogenous thyroid hormone is administered. In Graves' disease, the gland is driven by thyroid-stimulating immunoglobulin and is therefore nonsuppressible. Failure of radioiodine uptake to decrease or serum T_4 to fall after administration of triiodothyronine (liothyronine) in doses of 75-100 μg/d for 7-21 days is an indication that the gland is functioning autonomously and is strongly suggestive of Graves' disease. In addition, failure of TSH to rise following administration of TRH intravenously suggests that circulating levels of thyroid hormone are excessive and have suppressed the pituitary gland. The suppression test is useful in younger patients with ophthalmopathy who are euthyroid by other tests. The TRH test is very useful in older patients who present with cardiac symptoms and no striking elevation of circulating thyroid hormone levels. The suppression test is contraindicated in these older patients, because of the hazard of cardiac complications. Echography and CT scans of the orbit have revealed muscle enlargement in most patients with Graves' disease even when there is no clinical evidence of ophthalmopathy. In patients with clinical evidence of ophthalmopathy, orbital muscle enlargement is frequently striking.

Differential diagnosis

Graves' disease occasionally presents in an unusual or atypical fashion, in which case the diagnosis may not be obvious. Marked muscle atrophy may suggest severe myopathy that must be differentiated from primary neurologic disorder. Older patients with Graves' disease may present with symptoms of heart involvement, especially refractory atrial fibrillation or heart failure. This is characterized by high-output heart failure or chronic atrial fibrillation relatively insensitive to digoxin. Finally, patients with Graves' disease may present with infertility and amenorrhea as the primary symptoms. In all of these instances, the diagnosis of Graves' disease can usually be made with the above clinical and laboratory studies.

Complications

Thyrotoxic crisis ("*thyroid storm*") is the acute exacerbation of all of the symptoms of thyrotoxicosis, often presenting as a syndrome of such severity as to be life-threatening. Occasionally, thyroid storm may be mild and present simply as an unexplained febrile reaction after thyroid surgery in a patient who has been inadequately prepared. More commonly, it occurs in a more severe form, after surgery, radioactive iodine therapy, or parturition in a patient with thyrotoxicosis who has been inadequately controlled, or during a severe, stressful illness or disorder such as uncontrolled diabetes, trauma, acute infection, severe drug reaction, or myocardial infarction. The clinical manifestations of thyroid storm are marked hypermetabolism and excessive adrenergic response. Fever ranges from 38 to 41°C (100-106 °F) and is associated with flushing and sweating. There is marked tachycardia, often with atrial fibrillation, high pulse pressure, and occasionally heart failure. Central nervous system symptoms include marked agitation, restlessness, delirium, and coma. Gastrointestinal symptoms will include nausea, vomiting, diarrhea, and jaundice. A fatal outcome will be associated with heart failure and shock.

At one time it was thought that "dumping" of stored thyroxine and triiodothyronine was responsible for thyroid storm. Careful studies have revealed that the serum levels of T_4 and T_3 in patients with thyroid storm are not higher than in thyrotoxic patients without this condition. There is no evidence that thyroid storm is due to excessive production of triiodothyronine. There is evidence that in thyrotoxicosis there are increased numbers of binding sites for catecholamines, so that heart and nervous tissues have increased sensitivity to circulating catecholamines. In addition, there is decreased binding to TBG, with elevation of free T_3 and T_4. The present theory is that in this setting, with increased binding sites available for catecholamines, an acute illness, infection, or surgical stress triggers an outpouring of catecholamines which, in association with high levels of free T_4 and T_3, precipitate the acute problem.

The most striking clinical diagnostic feature of thyrotoxic crisis is hyperpyrexia out of proportion to other findings. Laboratory findings include elevated serum T_4, FT_4I, and T_3 by radioimmunoassay.

Treatment

Although autoimmune mechanisms are responsible for the syndrome of Graves' disease, management has been largely directed toward controlling the hyperthyroidism. Three good methods are available: (1) antithyroid drug therapy, (2) surgery, and (3) radioactive iodine therapy.

Antithyroid drug therapy

In general, antithyroid drug therapy is most useful in young patients with small glands and mild disease. The drugs (propylthiouracil or methimazole) are given until the disease undergoes spontaneous remission. This occurs in 20-40% of patients treated for 6 months to 15 years. Although this is the only therapy that leaves an intact thyroid gland, it does require a long period of observation, and the incidence of relapse is high, perhaps 60-80% even in selected patients. Antithyroid drug therapy is generally started with large divided doses, when the patient becomes clinically euthyroid, maintenance therapy may be achieved with a lower single morning dose. A common regimen consists of giving propylthiouracil, 100-150 mg every 6 hours initially, and then in 4-8 weeks reducing the dose to 50-200 mg once daily in the morning. Propylthiouracil has one advantage over methimazole in that it partially inhibits the conversion of T_4 to T_3, so that it is effective in bringing the levels of activated thyroid hormone down more quickly. On the other hand, methimazole has a longer duration of action and is more useful if a single daily dose is desirable. The laboratory tests of most value in monitoring the course of therapy are serum T_3 by RIA or the free thyroxine index.

1. *Duration of therapy*. The optimal duration of therapy of antithyroid drugs is quite variable. Sustained remissions can be predicted in patients with the following characteristics: (1) The thyroid gland returns to normal size. (2) The patient can be controlled on a relatively small dose of antithyroid drugs. (3) HLA typing shows an absence of HLA-B8. (4) TSI is no longer detectable in the serum. (5) The thyroid gland becomes normally suppressible following the administration of liothyronine.
2. *Reaction to drugs*. Reactions to antithyroid drugs most commonly involve either a rash (about 5%) or agranulocytosis (about 0.5%). The rash may frequently be managed by simple addition of antihistamines and is not necessarily an indication for discontinuing the medication unless it is severe and generalized. Agranulocytosis is an indication for immediate cessation of antithyroid drug therapy, institution of appropriate antibiotic therapy, and shifting to an alternative type of therapy. Agranulocytosis is usually heralded by sore throat and fever. Thus, all patients receiving antithyroid drugs are instructed that if sore throat or fever develops, they should immediately stop taking the drug and see a physician, who should obtain a white count and differential count. Cholestatic jaundice, hepatocellular toxicity, and acute arthralgia are rare side-effects but require cessation of drug therapy when they do occur.

Surgical treatment

Subtotal thyroidectomy is the treatment of choice for patients with very large glands or multinodular goiters. The patient is prepared with antithyroid drugs until euthyroid (about 6 weeks). In addition, starting 2 weeks before the day of operation, the patient is given saturated solution of potassium iodide, 5 drops twice daily. This regimen has been shown empirically to diminish the vascularity of the gland and to simplify surgery.

There is disagreement about how much thyroid tissue should be removed. Total thyroidectomy is usually not necessary unless the patient has severe progressive ophthalmopathy. On the other hand, if too much thyroid tissue is left behind, the disease will relapse. Most surgeons leave 2-3 g of thyroid tissue on either side of the neck. Although some patients do not require thyroid supplementation following thyroidectomy for Graves' disease, most patients do.

Hypoparathyroidism and recurrent laryngeal nerve injury occur as complications of surgery in about 1% of cases.

Radioactive iodine therapy

Therapy with sodium iodide I 131 is the preferred treatment for most patients over age 21. In many patients without underlying heart disease, radioactive iodine may be given immediately in a dosage of 80-120 μCi/g of thyroid weight estimated on the basis of physical examination and sodium iodide I 123 rectilinear scan. The dosage is corrected for iodine uptake according to the following formula:

$$80\text{-}120\ \mu\text{Ci/g} \times \begin{array}{c}\text{Estimated}\\ \text{weight of}\\ \text{gland in}\\ \text{grams}\end{array} \times \frac{100}{\begin{array}{c}\text{24-Hour ratio}\\ \text{active iodine}\\ \text{uptake (\%)}\end{array}} = \text{Dose in } \mu\text{Ci}$$

In patients with underlying heart disease, severe thyrotoxicosis, or large glands (over 100 g), it is often desirable to achieve a euthyroid state before radioactive iodine is started. These patients are treated with antithyroid drugs until they are euthyroid; medication is then stopped for 5-7 days; the radioactive iodine uptake is then determined and a scan is done; and a dose of 120-150 μCi/g of estimated thyroid weight is calculated on the basis of this uptake. A slightly larger dose is necessary in patients previously treated with antithyroid drugs. Following the administration of radioactive iodine, the gland will shrink and the patient will usually become euthyroid over a period of 6-12 weeks.

The major complication of radioactive therapy is hypothyroidism, which ultimately develops in 80% or more of patients who are adequately treated. This need not be considered a true complication and may indeed be the best assurance that the patient will not have a recurrence of hyperthyroidism. Serum free thyroxine index and TSH levels should be followed, and if they show the development of hypothyroidism, prompt replacement therapy with levothyroxine, 0.15--0.2 mg daily, is instituted.

It is of interest that hypothyroidism occurs after any type of therapy for Graves' disease and may indeed be the normal outcome of "burned-out" Graves' disease. Accordingly, all patients with Graves' disease require lifetime follow-up to be certain that they remain in a euthyroid state.

Other medical measures

During the acute phase of thyrotoxicosis, beta-adrenergic blocking agents are extremely helpful. Propranolol, 10-40 mg every 6 hours, will control tachycardia, hypertension, and atrial fibrillation. This drug is gradually withdrawn as serum thyroxine levels return to normal.

Adequate nutrition, including multivitamin supplements, is essential. Barbiturates accelerate T_4 metabolism, and phenobarbital may be helpful both for its sedative effect and to lower T_4 levels.

Treatment of complications

Thyrotoxic crisis

Thyrotoxic crisis (thyroid storm) requires vigorous management. Propranolol, 1-2 mg slowly intravenously or 40-80 mg every 6 hours orally, is extremely helpful in controlling the severe cardiovascular symptoms. In the presence of severe heart failure or asthma, one can use reserpine, 1 mg intramuscularly every 6 hours, or guanethidine, 1-2 mg/kg in divided doses orally daily. Hormone release is retarded by the administration of sodium iodide, 1 g intravenously over a 24-hour period, or saturated solution of potassium iodide, 10 drops twice daily. Hormone synthesis is blocked by the administration of propylthiouracil, 250 mg every 6 hours. The conversion of T_4 to T_3 is partially blocked by the combination of propranolol and propylthiouracil and the administration of hydrocortisone hemisuccinate, 50 mg intravenously every 6 hours. Supportive therapy includes a cooling blanket and acetaminophen to help control fever. Aspirin is probably contraindicated, because of its tendency to bind to TBG and displace thyroxine, rendering more thyroxine available in the free state. Fluids, electrolytes, and nutrition are important. For sedation, phenobarbital is probably best because it accelerates the peripheral metabolism and inactivation of thyroxine and triiodothyronine, ultimately bringing these levels down. Oxygen, diuretics, and digitalis are indicated for heart failure. Finally, it is essential to treat the underlying disease process that may have precipitated the acute exacerbation. Thus, antibiotics, anti-allergy drugs, and postoperative care are indicated for management of these problems. Extreme measures (rarely needed) to control thyrotoxic crisis include plasmapheresis to remove high levels of circulating thyronines, or peritoneal dialysis for the same purpose.

Ophthalmopathy

Management of ophthalmopathy due to Graves' disease involves treatment of thyroid disease, usually by total surgical excision of the thyroid gland or total ablation of the gland with radioactive iodine. Although there is controversy over the need for total ablation, removal or destruction of the thyroid gland certainly prevents exacerbations and relapses that will worsen residual ophthalmopathy. Keeping the head elevated will diminish periorbital edema. For the severe acute inflammatory reaction, a short course of corticosteroid therapy is frequently effective, eg, prednisone, 100 mg daily orally in divided doses for 7-14 days, then every other day for 6-12 weeks. If corticosteroid therapy is not effective, external x-ray therapy to the retro-orbital area may be helpful. The dose is usually 2000 R in 10 fractions given over a period of 2 weeks. The lens and anterior chamber structures must be shielded.

In very severe cases with threat to vision, orbital decompression can be used. One type of orbital decompression involves the transantral approach through the maxillary sinus, removing the floor and the lateral walls of the orbit. This has been extremely effective, and exophthalmos can be reduced by 5-7 mm in each eye by this technique. After the acute process has subsided, the patient is frequently left with double vision or lid abnormalities due to muscle fibrosis and contracture. These can be corrected by cosmetic lid surgery or eye muscle surgery.

Thyrotoxicosis and pregnancy

Thyrotoxicosis during pregnancy presents a special problem. Radioactive iodine is contraindicated, because it crosses the placenta freely and may injure the fetal thyroid. Two good alternatives are available. If the disease is detected during the first trimester, the patient can be prepared with propylthiouracil, and subtotal thyroidectomy can be performed safely during the mid trimester. It is essential to provide thyroid supplementation during the balance of the pregnancy. Alternatively, the patient can be treated with antithyroid drugs throughout the pregnancy, postponing the decision regarding long-term management until after delivery. The dosage of antithyroid drugs must be kept to the minimum necessary to control symptoms, because these drugs cross the placenta and may affect the function of the fetal thyroid gland. If the disease can be controlled by initial doses of propylthiouracil of 300 mg or less and maintenance doses of 50-150 mg/d, the likelihood of fetal hypothyroidism is extremely small. The free thyroxine index should be maintained in the upper range of normal by downward adjustment of propylthiouracil dosage. Supplemental thyroxine is not necessary. Breast feeding is not contraindicated, because propylthiouracil is not concentrated in the milk.

Graves' disease may occur in the newborn infant. There seem to be 2 neonatal forms of the disease. In both types, the mother has a current or recent history of Graves' disease. In the first type, the child is born small, with weak muscles, tachycardia, fever, and frequently respiratory distress or neonatal jaundice. Examination reveals an enlarged thyroid gland and occasionally prominent, puffy eyes. The heart rate is rapid, temperature is elevated, and heart failure may ensue. Laboratory studies reveal an elevated FT_4I, a markedly elevated T_3 and usually a low TSH—in contrast, to normal infants who have elevated TSH at birth. Bone age may be accelerated. TSI is often found in the serum of both the infant and the mother. The pathogenesis of this syndrome is thought to involve transplacental transfer of TSI from mother to fetus, with subsequent development of thyrotoxicosis. The disease is self-limited and subsides over a period of 4-12 weeks, coinciding with the fall in the child's TSI. Therapy includes propylthiouracil in a dose of 5-10 mg/kg/d (in divided doses at 8-hour intervals); strong, iodine (Lugol's) solution, 1 drop (8 mg potassium iodide) every 8 hours; and propranolol, 2 mg/kg/d in divided doses. In addition, adequate nutrition, antibiotics for infection if present, sedatives if necessary, and supportive therapy are indicated. If the child is very toxic, corticosteroid therapy (prednisone, 2 mg/kg/d) will partially block conversion of T_4 to T_3 and may be helpful in the acute phase. The above medications are gradually reduced as the child improves and can usually be discontinued by 6-12 weeks.

A second form of neonatal Graves' disease occurs in children from families with a high incidence of Graves' disease. Symptoms develop more slowly and may not be noted, until the child is 3-6 months old. This syndrome is thought to be a true genetic inheritance of defective lymphocyte immunoregulation. It is much more serious in terms of severity, with a 20% mortality rate and evidence of persistent brain dysfunction even after successful treatment. The hyperthyroidism may persist for months or years and requires prolonged therapy.

Course and prognosis

In general, the course of Graves' is one of remissions and exacerbations over a protracted period of time unless the gland is destroyed by surgery or radioactive iodine. Although some patients may remain euthyroid for long periods after treatment, many eventually develop hypothyroidism. Lifetime follow-up is therefore indicated for all patients with Graves' disease.

Other Forms of Thyrotoxicosis

Toxic adenoma (Plummer's disease)

Thyrotoxicosis may develop because of a single adenoma, usually a colloid adenoma, hypersecreting T_3 and T_4. These lesions start out as a "hot nodule" on the thyroid scan, slowly increase in size, and gradually suppress the other lobe of the thyroid gland. The typical patient is an older individual (usually over 40) who has noted recent growth of a long-standing thyroid nodule. Symptoms of weight loss, weakness, shortness of breath, palpitation, tachycardia, and heat intolerance are noted. Eye signs are almost never present. Physical examination reveals a definite nodule on one side, with very little thyroid tissue on the other side. Laboratory studies usually reveal marked elevation in serum T_3 levels, with only borderline elevation of thyroxine levels. The scan reveals that the nodule is "hot." Toxic adenomas are almost always colloid adenomas and almost never malignant. They are easily managed by administration of antithyroid drugs such as propylthiouracil, 100 mg every 6 hours, or methimazole, 10 mg every 6 hours, followed by unilateral lobectomy; or with radioactive iodine. Sodium iodide 131 I in doses of 20-30 mCi is usually required to destroy the benign neoplasm. Radioactive iodine is preferable for smaller toxic nodules, but larger ones are best managed by operation.

Toxic multinodular goiter (Marine-lenhart syndrome)

This disorder occurs in older patients with long-standing multinodular goiter. Ophthalmopathy is extremely rare. Clinically, the patient presents with tachycardia, heart failure, or arrhythmia and sometimes weight loss, nervousness, tremors, and sweating. Laboratory studies reveal a striking elevation in serum T_3 levels, with less striking elevation in serum T_4 levels. Radioiodine scan reveals multiple functioning nodules in the gland, although occasionally one may see an irregular, patchy distribution of radioactive iodine.

Hyperthyroidism in patients with multinodular goiters can often be precipitated by the administration of iodides (jodbasedow effect, or iodide-induced hyperthyroidism). It was at first thought that jodbasedow effect occurred only in iodine-deficient geographic areas and that administration of iodides supplied the "building blocks" for hypersecretion of T_4 and T_3. However, Vagenakis and co-workers have reported that 4 out of 8 patients with nontoxic goiters who lived in a high-iodine area and who received 5 drops of saturated solution of potassium iodide daily developed hyperthyroidism. Thus, large doses of iodide can induce hyperthyroidism in any patient with a nontoxic goiter regardless of the geographic locality. The management of this problem is somewhat difficult. Ideally, subtotal thyroidectomy would be the treatment of choice, but these patients are often poor operative risks. Alternatively, one can give relatively large doses of radioactive iodine, but the multinodular goiter will not be destroyed, and repeated doses of radioactive iodine may be necessary. Antithyroid drugs are effective only on a temporary basis to bring the patient to a euthyroid state. They should be used prior to surgery or radioactive iodine therapy in order to diminish the stress of these procedures. Propranolol may be helpful to control the cardiac problems.

Subacute thyroiditis

This entity will be discussed in a separate section, but it should be mentioned here that thyroiditis may present as an acute release of T_4 and T_3 with symptoms of thyrotoxicosis. The typical patient is a young person with a history of malaise and complaints of neck pain, tachycardia, weight loss, and sweating. The thyroid gland is large and slightly tender. Laboratory findings reveal elevated free thyroxine index and serum T_3 but a very low radioactive iodine uptake, often less than 1%. "The erythrocyte sedimentation rate is usually markedly elevated. This disease will usually subside spontaneously, although the patient may require propranolol to control cardiovascular symptoms and aspirin for local pain and discomfort.

Hashimoto's thyroiditis

Hashimoto's thyroiditis can also go through an acute phase with increased T_3 and T_4 release, producing transient hyperthyroidism. This too is associated with high T_3 and T_4 levels and very low radioactive iodine uptake. It is important to differentiate Hashimoto's thyroiditis from true Graves' disease, since the former will resolve spontaneously and does not require surgery, antithyroid drug therapy, or radioactive iodine. In most cases, propranolol is the only therapy required.

Thyrotoxicosis factitia

This is a psychoneurotic disturbance in which excessive amounts of thyroxine or thyroid hormone are ingested, usually for purposes of weight control. The individual is often someone connected with medicine to whom thyroid medication is easily available. Features of thyrotoxicosis, including weight loss, nervousness, palpitation, tachycardia, and tremor, may be present, but no goiter or eye signs. Characteristically, the serum T_4 and T_3 levels are elevated and radioactive iodine uptake nil. Management requires careful discussion of the hazards of long-term thyroxine therapy, particularly cardiovascular damage and muscle wasting. Formal psychotherapy may be necessary.

Rare forms of thyrotoxicosis

Struma ovarii

This is a syndrome in which teratoma of the ovary develops that contains thyroid tissue, and the thyroid tissue becomes hyperactive.

Mild features of thyrotoxicosis result, such as weight loss and tachycardia, but no goiter or eye signs. Free thyroxine index and serum T_3 are usually mildly elevated, and the radioiodine uptake in the neck will be nil. Body scan reveals uptake of radioiodine in the pelvis. The disease is curable with removal of the teratoma.

Thyroid carcinoma

Carcinoma of the thyroid, particularly follicular carcinoma, may concentrate radioactive iodine, but only rarely does it retain the ability to convert this iodide into active hormone. There have been a few instances of metastatic thyroid cancer presenting with hyperthyroidism. The clinical picture consists of weakness, weight loss, and palpitation, without goiter or ophthalmopathy. Body scan reveals areas of uptake usually distant from the thyroid, eg, bone or lung. Treatment with large doses of radioactive iodine may destroy the metastatic deposits.

Hydatidiform mole

Hydatidiform mole produces chorionic gonadotropin, which has intrinsic TSH-like activity. This may induce thyroid hyperplasia, increased iodine turnover, and mild elevation of serum T_4 and T_3 levels. It is rarely associated with overt thyrotoxicosis and is totally curable by removal of the mole.

Syndrome of inappropriate TSH secretion

A group of patients have recently been reported with elevated serum immunoreactive TSH in association with elevated serum free thyroxine values. This has been called the "syndrome of inappropriate TSH secretion." Two types of problems are found: (1) TSH secreting pituitary adenoma and (2) nonneoplastic pituitary hypersecretion of TSH.

Patients with TSH-secreting pituitary adenomas usually present with mild thyrotoxicosis and goiter, usually with amenorrhea. There are no eye signs of Graves' disease. Study reveals elevated total and free serum T_4 and T_3. There is no response to TRH, and the increased radioactive iodine uptake is not suppressible with exogenous thyroid hormone. On the other hand, serum TSH, usually undetectable in Graves' disease, is elevated. Visual field examination may reveal temporal defects, and CT scan of the sella reveals a pituitary tumor. Management usually involves control of the thyrotoxicosis with antithyroid drugs and removal of the pituitary tumor via transsphenoidal hypophysectomy. If the tumor cannot be completely removed, it may be necessary to treat residual tumor with radiation therapy and to control thyrotoxicosis with radioactive iodine.

Nonneoplastic pituitary hypersecretion of TSH is essentially a form of pituitary (and occasionally peripheral) resistance to T_3 and T_4. Thus, increasing levels of T_3 and T_4 fail to suppress pituitary TSH secretion, and TSH pours out. Patients with only pituitary resistance to T_3 and T_4 present with goiter and hyperthyroidism, elevated free T_4, elevated TSH, normal response to TRH, and a normal response to T_3 suppression. These patients act as if the pituitary set point for T_3 suppression is at a higher than normal level. Patients with both pituitary and peripheral resistance to thyroid hormones will present with goiter, elevated T_3 and T_4, and elevated TSH, but they are euthyroid or mildly hypothyroid. This syndrome is familial and has been called *Refetoff's syndrome*.

Management of these syndromes is difficult. If the patient is hypothyroid or euthyroid, T_3 therapy may be effective in maintaining a euthyroid state and reducing TSH. In the hyperthyroid patient, bromocriptine may lower TSH and control thyrotoxic symptoms.

Nontoxic Goiter

Etiology

Nontoxic goiter usually represents enlargement of the thyroid gland from TSH stimulation, which in turn results from inadequate thyroid hormone synthesis. Iodine deficiency was the most common cause of nontoxic goiter or "endemic goiter", with the widespread use of iodized salt and the introduction of iodides into fertilizers, animal feeds, and food preservatives, iodide deficiency in developed countries is relatively rare. The only remaining areas of endemic goiter are in inland, undeveloped regions such as the Himalayas, the Andes, Central Africa, and New Guinea. It does not exist in the USA. Estimated optimal iodine requirements for adults are in the range of 150-300 μg/d. In endemic goiter areas, the daily urinary excretion of iodine falls below 50 μg/d; in areas where iodine is extremely scarce, excretion falls below 20 μg/d. The mechanism for goiter formation is thought to be low iodine intake, impaired hormone formation, increased TSH secretion, and hyperplasia of thyroid cells.

Dietary goitrogens are a rare cause of goiter, and of these the most common is iodide itself. Large doses of iodides used (for example) in the treatment of chronic pulmonary disease may in susceptible individuals produce goiter and hypothyroidism. Withdrawal of iodide reverses the process. Other goitrogens include lithium carbonate, used for the treatment of manic-depressive psychosis, and some vegetable foodstuffs such as thioglucosides, found in cabbage, and goitrin, found in certain roots and seeds. Cyanogenic glycosides found in cassava can release thiocyanates that may cause goiter, particularly in the presence of iodide deficiency. The role of these vegetable goitrogens in the production of goiter is not clearly established.

Thyroid enlargement may occur in adolescents, particularly at the time of menarche, or as a result of pregnancy or ingestion of or contraceptive drugs. It was at first thought that these goiters were du to increased thyroxine-binding globulin and increased demands upon the thyroid to maintain a larger extrathyroidal hormone pool. However, many of these patients have underlying chronic lymphocytic thyroiditis (Hashimoto's thyroiditis), which is the real cause of the goiter. Hashimoto's thyroiditis is a common illness, particularly in women. Hormone synthesis is impaired, and thyroid enlargement develops. In subacute thyroiditis, thyroid enlargement occurs in association with acute inflammatory changes, presumably due to viral infection.

Nontoxic goiter may occur as a result of synthesis resulting from genetic deficiencies in enzymes necessary for hormone biosynthesis (thyroid dyshormonogenesis). These effects may be complete, resulting in a syndrome of cretinism with goiter; or partial, resulting in nontoxic goiter with mild hypothyroidism. At least 5 separate biosynthetic abnormalities have been reported: (1) impaired transport of iodine: (2) deficient peroxidase with impaired oxidation of iodide to iodine and failure to incorporate iodine into thyroglobulin; (3) impaired coupling of iodinated tyrosines to triiodothyronine or tetraiodothyronine: (4) absence or deficiency of iodotyrosine deiodinase, so that iodine is not conserved within the gland; and (5) excessive production of metabolically inactive iodoprotein by the thyroid gland. In addition, it has been demonstrated that impaired formation of thyroglobulin, or synthesis of abnormal thyroglobulin, may result in inadequate hormone production. In all of these syndromes, impaired production of thyroid hormones presumably results in TSH release and goiter formation.

Finally, thyroid enlargement can be due to a benign lesion, such as adenoma, or a malignant one such as carcinoma.

Pathogenesis

It has been demonstrated quite clearly that the pathogenetic mechanism of nontoxic goiter due to dyshormonogenesis or Hashimoto's thyroiditis involves first diffuse hyperplasia, followed by focal hyperplasia with necrosis and hemorrhage, and finally the development of new areas of focal hyperplasia. Thus, a diffuse nontoxic goiter progresses over a period of time to a multinodular nontoxic goiter.

Clinical Findings

Symptoms and signs

Patients with nontoxic goiter usually present with thyroid enlargement, which may be diffuse or multinodular. The gland may be relatively firm but is often extremely soft. Over a period of time, the gland becomes larger and larger, so that in long-standing multinodular goiter, huge goiters may develop and extend inferiorly to present as substernal goiter. The patient may complain of pressure symptoms in the neck, particularly on moving the head upward or downward, and of difficulty in swallowing. Vocal cord paresis is rare. There may be symptoms of mild hypothyroidism, but most of these patients are euthyroid. Thyroid enlargement represents compensated hypothyroidism.

Laboratory findings

Laboratory studies will reveal a low or normal free thyroxine index and, usually, normal levels of TSH. It is assumed that at the time the studies are made the patient has come into equilibrium or balance, so that TSH levels are not elevated. In patients with dyshormonogenesis due to abnormal iodoprotein synthesis, PBI may be elevated out of proportion to serum T_4, because of secretion of nonhormonal organic iodide compounds.

Scan studies

Isotope scan usually reveals a patchy uptake, frequently with focal areas of uptake corresponding to “hot” nodules. Radioactive uptake usually is suppressible on administration of thyroid hormones such as liothyronine. Echography may reveal cystic changes in one or more of the nodules, representing previous hemorrhage and necrosis.

Differential Diagnosis

The major problem of differential diagnosis is to rule out cancer. This will be discussed in more detail in the section on thyroid carcinoma.

Treatment

With the exception of those due to neoplasm, the management of nontoxic goiter consists simply of giving thyroid hormones until TSH is completely suppressed. Levothyroxine in doses of 0.15-0.2 mg daily will suppress pituitary TSH and result in slow regression of the goiter as well as correction of hypothyroidism. Long-standing goiters may have areas of necrosis, hemorrhage, and scarring that will not regress on thyroxine therapy. However, the lesions will not grow while the patient is taking thyroxine.

Note that the left lobe of the gland extends from the middle of the thyroid cartilage to just above the clavicle. The pressure of this enlargement has caused deviation of the trachea to the right. The surface of the gland is irregular, with many large and small nodules Although these multinodular goiters are rarely malignant, the size of the mass with resulting pressure symptoms may require subtotal thyroidectomy.

Course and Prognosis

Patients with nontoxic goiter must usually take levothyroxine for life. They should avoid iodides, which may induce either hyperthyroidism or, in the absence of thyroxine therapy, hypothyroidism. Occasionally, single adenomas or several adenomas will become hyperplastic and produce toxic nodular goiter. Nontoxic goiter is often familial, and other members of the family should be examined and observed for the possible development of goiter.

Surgery is not necessary for nontoxic goiter unless obstructive symptoms develop or marked substernal extension occurs. Substernal goiter does not respond to levothyroxine therapy and must be removed surgically.

Thyroiditis

Subacute Thyroiditis

Subacute thyroiditis (De Quervain's thyroiditis, granulomatous thyroiditis) is an acute inflammatory disorder of the thyroid gland most likely due to viral infection. A number of viruses, including mumps virus, coxsackievirus, and adenoviruses, have been implicated, either by finding the virus in biopsy specimens taken from the gland or by demonstration of rising titers of antiviral antibodies in the blood during the course of the infection. Pathologic examination reveals moderate thyroid enlargement and a mild inflammatory reaction involving the capsule. Histologic features include destruction of thyroid parenchyma and the presence of many large phagocytic cells, including giant cells.

Clinical findings

Symptoms and signs

Subacute thyroiditis usually presents with fever, malaise, and soreness in the neck, which may extend up to the angle of the jaw or toward the ear lobes on one or both sides of the neck. Initially, the patient may have symptoms of hyperthyroidism with palpitations, agitation, and sweats. There is no ophthalmopathy. On physical

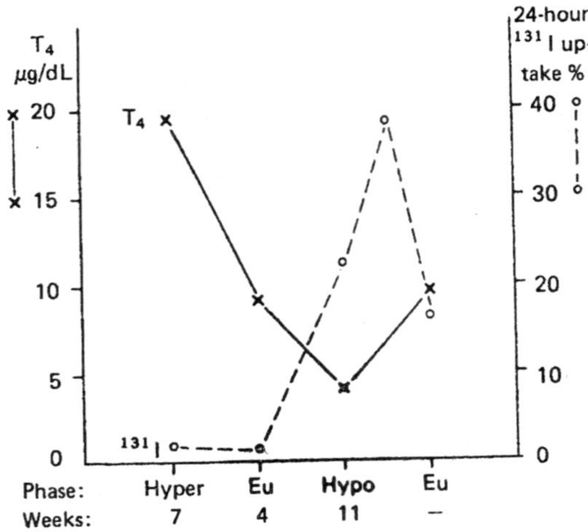

Fig. 7.27. Changes in serum T_4 and radioactive iodine uptake in patients with subacute thyroiditis.

examination, the gland is exquisitely tender, so that the patient will object to pressure upon it. There are no signs of local redness or heat suggestive of abscess formation. Clinical signs of toxicity, including tachycardia, tremor, and hyperreflexia, may be present. Laboratory studies will vary with the course of the disease. Initially, T_3 by RIA, T_4 by RIA, and resin T_3 uptake are usually elevated, whereas serum TSH and thyroid radioactive iodine uptake are extremely low. The erythrocyte sedimentation rate is markedly elevated, sometimes as high as 100 mm/h by the Westergren scale. Antithyroid antibodies are usually not detectable in serum. As the disease progresses, T_4 and T_3 will drop, TSH will rise, and symptoms of hypothyroidism are noted. Later, radioactive iodine uptake will rise, reflecting recovery of the gland from the acute insult.

Differential diagnosis

Subacute thyroiditis can be differentiated from other viral illnesses by the involvement of the thyroid gland. It is differentiated from Graves' disease by the presence of low thyroid radioiodine uptake at a time when T_3 and T_4 levels are elevated.

Course

Subacute thyroiditis usually resolves spontaneously over weeks or months. Occasionally, the disease may begin to resolve and then suddenly get worse, sometimes involving first one lobe of the thyroid gland and then the other. Exacerbations come at a time when the T_4 levels have fallen, TSH has risen, and the gland is starting to recover function. Rarely, the course may extend over several years, with repeated bouts of inflammatory disease.

Treatment

In most cases, only symptomatic treatment is necessary, eg, aspirin, 0.6 g 4 times daily. If pain, fever, and malaise are disabling, a short course of corticosteroids such as prednisone, 20 mg 3 times daily for 7-10 days, may be necessary to reduce the inflammation. Therapy with thyroid hormones, either levothyronine, 25 μg 3 times daily, or levothyroxine, 0.2 mg once daily, may be necessary during the hypothyroid phase of the illness in order to prevent reexacerabation of the disease induced by the rising TSH levels. In most patients, complete recovery occurs, but in about 10% of cases permanent hypothyroidism ensues and long-term thyroxine therapy is necessary.

Chronic (Hashimoto's) Thyroiditis

Chronic thyroiditis (Hashimoto's thyroiditis, lymphocytic thyroiditis) is probably the most common cause of hypothyroidism and goiter in the USA. It is certainly the major cause of goiter in children and in young adults and is probably the major cause of "idiopathic myxedema," which represents an end stage of Hashimoto's thyroiditis with total atrophy of the gland. *Riedel's struma* is probably a variant of Hashimoto's thyroiditis with extensive fibrosis extending outside of the gland and involving overlying muscle and surrounding tissues. Riedel's struma presents as a stony hard mass that must be differentiated from thyroid cancer.

Etiology and pathology

Hashimoto's thyroiditis is thought to be an autoimmune disease in which there is a defect in "suppressor" T cells, and "helper" T cells stimulate B lymphocytes to produce antithyroid antibodies, including antimicrosomal and antithyroglobulin antibodies. During the early phases of Hashimoto's thyroiditis, antithyroglobulin antibodies are markedly elevated, and antimicrosomal antibodies are not so high. Later, antithyroglobulin antibodies may disappear, but the antimicrosomal antibodies will be present for many years. Pathologically, there is massive infiltration of lymphocytes into the thyroid gland, with total destruction of normal thyroidal architecture. There, may be actual formation of lymphoid follicles and germinal centers. The follicular epithelial cells are frequently enlarged and contain a basophilic cytoplasm (Hurthle cells). Destruction of the gland results in a fall in serum T_3 and T_4 levels and a rise in TSH: Initially, TSH may maintain adequate hormonal synthesis by the development of thyroid enlargement or goiter, but in many cases the gland fails, and hypothyroidism with or without goiter ensues.

Hashimoto's thyroiditis is part of a spectrum of thyroid diseases that includes Graves' disease at one end and idiopathic myxedema at the other. It is familial and may be associated with other autoimmune diseases such as pernicious anemia, adrenocortical insufficiency, idiopathic hypoparathyroidism, myasthenia gravis, and vitiligo. *Schmidt's syndrome* consists of Hashimoto's thyroiditis, idiopathic adrenal insufficiency, hypoparathyroidism, diabetes mellitus, ovarian insufficiency (rarely), and candidal infections. Schmidt's syndrome represents destruction of multiple endocrine glands on an autoimmune basis.

Clinical findings

Symptoms and signs

Hashimoto's thyroiditis usually presents with goiter and mild hypothyroidism. The sex distribution is about 4 females to one male. The process is painless, and the patient may be unaware of the goiter unless it becomes very large.

Laboratory findings

There are multiple defects in iodine metabolism. Peroxidase activity is decreased, so that organification of iodine is impaired. This can be demonstrated by a positive perchlorate discharge test. In addition, iodination of metabolically inactive protein material occurs, so that there will be a disproportionately high serum PBI compared to serum T_4. Radioiodine uptake may be high, normal, or low. Circulating thyroid hormone levels are usually normal or low, and if low, TSH will be elevated.

The most striking laboratory finding is the high titer of antithyroid antibodies in the serum. Antithyroglobulin and antimicrosomal antibody tests are strongly positive in most patients with Hashimoto's thyroiditis.

An additional diagnostic test that has recently become available is the fine needle aspiration biopsy. With this technique, biopsy samples will reveal a heavy infiltration of lymphocytes as well as the presence of Hurthle cells on the smear.

Differential diagnosis

Hashimoto's thyroiditis must be differentiated from other causes of nontoxic goiter. This is best done by antibody studies and if necessary by fine needle aspiration biopsy.

Complications and sequelae

The major complication of Hashimoto's thyroiditis is progressive hypothyroidism. Rarely, a patient with Hashimoto's thyroiditis may develop lymphoma of the thyroid gland, but whether the 2 conditions are causally related is not clear. Thyroid lymphoma is characterized by rapid growth of the gland despite continued thyroid hormone therapy; the diagnosis of lymphoma must be made by surgical biopsy.

There is no evidence that adenocarcinoma of the thyroid gland occurs more frequently in patients with Hashimoto's thyroiditis, but the 2 diseases—chronic thyroiditis and carcinoma—can coexist in the same gland. Cancer must be suspected when a solitary nodule or thyroid mass grows or fails to regress while the patient is receiving maximal tolerated doses of thyroxine. Fine needle aspiration biopsy may be helpful in differential diagnosis.

Treatment

Treatment consists of giving levothyroxine, 0.2-0.3 mg daily. These doses are slightly higher than physiologic replacement in an effort to suppress TSH and allow regression of the goiter. Surgery is rarely indicated.

Course and prognosis

Without treatment, Hashimoto's thyroiditis often progresses from goiter and hypothyroidism to myxedema. The myxedema is totally prevented by adequate thyroxine therapy. Because Hashimoto's thyroiditis may be a part of a syndrome of multiple autoimmune diseases, the patient should, be followed for other illnesses such as pernicious anemia., adrenal insufficiency, hypothyroidism, or diabetes mellitus.

Hashimoto's thyroiditis may go through periods of activity when large amounts of T_4 and T_3 are released or "dumped," resulting in symptoms of thyrotoxicosis. This syndrome, which has been called spontaneously resolving hyperthyroidism, is characterized by low radioiodine uptake. However, it can be differentiated from subacute thyroiditis in that the gland is not tender, the erythrocyte sedimentation rate is not elevated, antithyroid antibodies are strongly positive, and fine needle aspiration biopsy reveals lymphocytes and Hurthle cells. Therapy is symptomatic, usually requiring only propranolol, until symptoms subside; T_4 supplementation may then be necessary.

Since Hashimoto's thyroiditis is part of a spectrum of autoimmune thyroid disease that includes Graves' disease, patients with Hashimoto's thyroiditis may develop true Graves' disease, occasionally with severe ophthalmopathy or dermopathy. The chronic thyroiditis may blunt the severity of the thyrotoxicosis, so that the patient may present with eye or skin complications of Graves' disease without marked thyrotoxicosis, a syndrome often called euthyroid Graves' disease. The thyroid gland will invariably be nonsuppressible, and this, plus the presence of antithyroid antibodies, will help to make the diagnosis. The ophthalmopathy and dermopathy are treated as if thyrotoxic Graves' disease were present.

Other Forms of Thyroiditis

The thyroid gland may be subject to acute abscess formation in patients with septicemia or acute bacterial endocarditis. Abscesses cause symptoms of pyogenic infection, with local pain and tenderness, swelling, and warmth and redness of the overlying skin. Needle aspiration may be helpful to confirm the diagnosis and identify the organism. Treatment includes antibiotic therapy and occasionally incision and drainage. A thyroglossal duct cyst may become infected and present as acute suppurative thyroiditis. This too will respond to antibiotic therapy and occasionally incision and drainage.

Ionizing radiation can induce both acute and chronic thyroiditis. Thyroiditis may occur acutely in patients treated with large doses of radioiodine and may be associated with release of thyroid hormones and an acute thyrotoxic crisis. Such an occurrence is extremely rare, however; the author has seen only 2 cases in several thousand patients treated for Graves' disease with radioactive iodine.

External radiation was used many years ago for the treatment of benign conditions such as severe acne and chronic tonsillitis or adenoiditis. This treatment was frequently associated with the development of focal thyroiditis and at times with goiter and hypothyroidism. If cancer can be ruled out by physical examination and other studies, these patients should be treated with long-term thyroxine therapy as described above in the discussion of Hashimoto's thyroiditis.

Effects of Acute and Chronic Illness of Thyroid Function

The effects of thyroid hormone are controlled by a number of factors, including the following: (1) T_3 and T_4 feedback on TSH secretion; (2) autoregulation of hormone synthesis, depending on iodide availability; (3) peripheral metabolism of T_4, with activation by conversion to T_3 or inactivation by conversion to rT_3; and (4) the binding equilibrium between serum thyroid-binding proteins and tissue receptors. Acute and chronic illnesses have striking effects on these systems, particularly the peripheral metabolism of T_4 to T_3.

Activation of the inner ring deiodinase accelerates conversion of T_4 to rT_3 and conversion of T_3 to 3,3′-T_2, markedly lowering the circulating level of T_3. This occurs physiologically in the fetus and pathologically in circumstances of carbohydrate restriction, as in malnutrition, starvation, anorexia nervosa, and diabetes mellitus. Among the drugs that tend to lower the circulating levels of T_3, corticosteroids and iodinated dyes are the most effective and propylthiouracil and propranolol are relatively weak. Finally, acute illness such as myocardial infarction, febrile illness, trauma, burns, and surgery, as well as chronic illness—especially chronic liver disease, chronic renal disease, and cancer in advanced stages—also inhibit conversion of T_4 to T_3. These conditions result in what has been called the low T_3 syndrome.

Low T_3 syndrome is characterized by normal or slightly elevated T_4, low T_3, increased rT_3, and a normal TSH. Data from normal, "sick," and hypothyroid individuals are compared. The "sick" patients have relatively normal serum T_4, compared with true hypothyroid patients, who have a very low T_4. T_3 is low in both "sick" and hypothyroid patients, but rT_3 is elevated in the "sick" group in comparison with both the normal and the hypothyroid groups. TSH is markedly elevated in the hypothyroid patients and normal in the "sick" patients. Further evidence that the "sick" patients are not hypothyroid

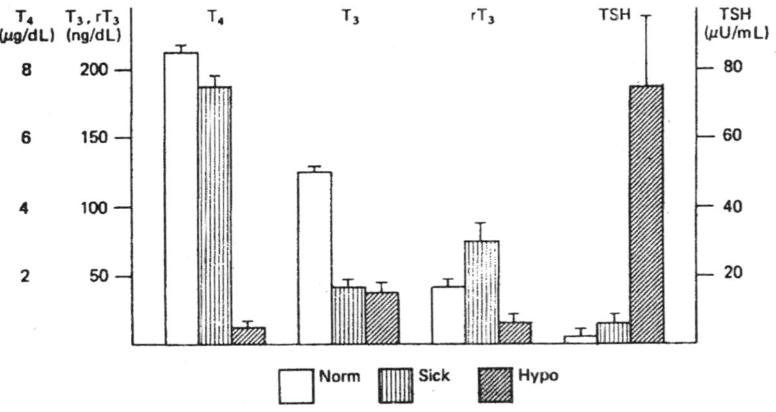

Fig. 7.28. Low T_3 syndrome.

can be obtained by the TRH test, which is normal in "sick" patients and hyperresponsive in hypothyroid patients.

There are 2 important areas of differential diagnosis in this syndrome: (1) Hypothyroidism must be ruled out, since failure to treat hypothyroidism in the presence of other illness could be catastrophic. The normal T_4, elevated rT_3, and normal TSH will rule out hypothyroidism. (2) Occasionally, T_4 may be slightly elevated, perhaps to compensate for the low T_3. These patients may be confused with hyperthyroid patients but can be differentiated by the low T_3 and a normal TRH test.

A group of patients have been reported with a *low T_3, low T_4, syndrome*. Generally, they are much sicker and are usually in the intensive care unit. Indeed, the presence of both low T_3 and low T_4 has been associated with high mortality rates and poor outcomes. This syndrome is characterized by low T_4, low T_3, high rT_3, normal TSH, and usually an impaired response to TRH. Note that T_4 and T_3 are low, but reverse T_3 is markedly elevated, and TSH is normal, in contrast to the hypothyroid patients. It is not clear why these patients are not hypothyroid. A possible explanation has been suggested by Chopra (1979), who noted that although the T_4 level was low, a greater percentage of T_4 was dialyzable, and the actual level of free T_4 was increased. This was not detected by the free thyroxine index, only by dialysis measurements. Chopra postulates an inhibitor of T_4-binding to thyroid-binding proteins to account for these findings.

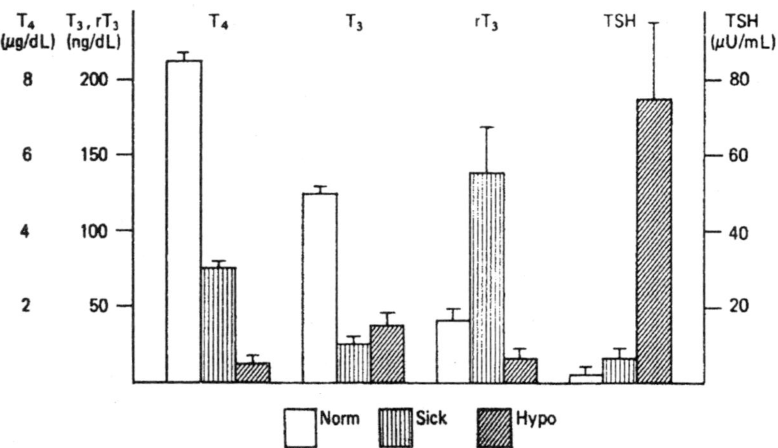

Fig. 7.29. Low T_3, low T_4 syndrome. Note that very sick patients have low T_4, low T_3, elevated rT_3, and normal TSH, compared with normal hypothyroid patients.

Clinical Implications of Low T_3 Syndrome

The clinical implications of these observations are important. The presence of low T_3 in milder diseases and low T_4 and low T_3 in very severe diseases suggests that these findings are related to the severity of the disease process. It is essential to rule out hyperthyroidism in low T_3 syndrome with elevated T_4 and hypothyroidism in both syndromes. Measurements of rT_3, TSH, and TRH response may be necessary. Finally, the FT_4I may not adequately reflect the level of free T_4 in severely ill patients, and direct determination of free T_4 may be necessary to clarify the thyroid status.

Management of Low T_3 Syndromes

Patients with these syndromes should not be treated with thyroid hormones to restore normal blood levels, since there is no evidence that treatment with T_3 or T_4 will be helpful and there are many reasons why such treatment might be hazardous. It is thought that these aberrations of thyroid hormone metabolism are an adaptive mechanism that may in some way protect the organism from excessive metabolic activity in the presence of serious illness. Certainly, one should correct any caloric or carbohydrate deprivation and treat the underlying illness, which will usually be accompanied by a return of thyroid hormone metabolism to normal.

Thyroid Nodules and Thyroid Cancer

In 95% of cases, thyroid cancer presents as a nodule or lump in the thyroid. In occasional instances, enlarged cervical lymph nodes will be the first sign of the disease. The latter is particularly apt to be the case in children, although on careful examination a small primary focus in the form of a thyroid nodule can usually be felt. Rarely, distant metastasis in lung or bone is the first sign of thyroid cancer.

Thyroid nodules are extremely common, particularly among women. The prevalence of thyroid nodules in the USA has been estimated to be about 4% of the adult population, in young children, the incidence is less than 1%: in persons aged 11-18 years, about 1.5%: and in persons over age 60, about 5%. In contrast to thyroid nodules, thyroid cancer is a rare condition—0.004% per year according to the Third National Cancer Survey. Thus, most thyroid nodules are benign, and a method must be utilized for identifying those that are likely to be malignant.

Etiology of Benign Thyroid Nodules

Benign conditions include focal areas of chronic thyroiditis, a dominant portion of a multinodular goiter, or a cyst involving either thyroid tissue, parathyroid tissue, or thyroglossal duct remnants. Other benign conditions include agenesis of one lobe of the thyroid, with hypertrophy of the other lobe presenting as a mass in the neck. It is usually the left lobe of the thyroid that fails to develop, and the hypertrophy occurs in the right lobe. Scarring in the gland following surgery—or regrowth of the gland after surgery or radioiodine therapy—can present with nodularity. Finally, benign neoplasms in the thyroid include follicular adenomas such as colloid or rnacrofollicular adenomas, fetal adenomas, embryonal adenomas, and Hurthle cell or oxyphil adenomas. Rare types of benign lesions include teratomas, lipomas, and hemangiomas. With the exception of thyroid hyperplasia of the right lobe of the gland in the presence of agenesis of the left lobe—and some follicular adenomas—all of the above lesions present as "cold" nodules on isotope scan.

Differentiation of Benign and Malignant Lesions of the Thyroid Gland

Risk factors that predispose to benign or malignant disease are set forth and discussed below.

History

A family history of goiter suggests benign disease, as does residence in an area of endemic goiter. On the other hand, a family history of medullary carcinoma, a history of recent growth, or a history of hoarseness, dysphagia, or obstruction strongly suggests cancer.

Probably the most important historical feature is exposure to ionizing radiation. As little as 6.5 rads to the thyroid gland received during the radiation treatment of tinea capitis has been reported to cause cancer in 0.11% of exposed children; the incidence of thyroid cancer in sibling controls was 0.02%. Radiation therapy to the thymus was used many years ago for the treatment of respiratory problems or poor growth in infants. The thyroid received dosages of 100-400 rads, and the incidence of thyroid cancer attributed to this source ranged from 0.5 to 5%. X-ray therapy to the neck and chest given to children or adolescents for acne, tonsillitis, adenoiditis, otitis media, or skin lesions with thyroid doses ranging from 200 to 1500 rads produced nodular goiter with an incidence of around 27% and thyroid cancer with an incidence of 5-7%. These tumors developed 10-40 years after radiation was administered, with a peak incidence at 20-30 years. Radiation fallout with a thyroid dose of 700-1400 rads has produced nodular goiter in approximately 40% of patients exposed and thyroid cancer in about 6%. On the other hand, radioiodine therapy, which exposes the thyroid to a dosage of around 10,000 rads, was rarely associated with the development of thyroid cancer, presumably because the thyroid gland is largely destroyed by these doses of radioiodine so that—although the incidence of postradiation hypothyroidism is high—the incidence of thyroid malignancy is extremely low.

Ninety percent of patients with radiation-induced thyroid cancer develop papillary carcinoma; the remainder develop follicular carcinoma. Medullary carcinoma and anaplastic carcinomas have been rare following radiation exposure. Data from several large series suggest that the incidence of cancer in a patient who presents with a solitary cold nodule of the thyroid gland and a history of therapeutic radiation of the head, neck, or chest is around 50%.

Physical characteristics

Physical characteristics associated with a low risk for thyroid cancer include older age, female sex, soft thyroid nodules, and the presence of a multinodular goiter. Individuals at higher risk for thyroid cancer include children, young adults, and males. A solitary firm or dominant nodule that is clearly different from the rest of the gland increases the risk of malignancy. Vocal cord paralysis, enlarged lymph nodes, and suspected metastases are strongly suggestive of malignancy.

Serum factors

A high titer of antithyroid antibodies in serum is associated with a low risk of thyroid cancer and suggests chronic thyroiditis as the cause of thyroid enlargement. On the other hand, an elevated serum calcitonin, particularly in patients with a family history of medullary carcinoma, strongly suggests the presence of thyroid cancer. Elevated serum thyroglobulin titers may be associated with metastatic follicular or papillary thyroid carcinoma but are usually not helpful in determining the nature of the thyroid nodule.

Scanning techniques

Scanning procedures can be used to identify "hot" or "cold" nodules, ie, those that take up more or less radioactive iodine than surrounding tissue. Hot nodules are almost never malignant, whereas cold ones may be. Scintillation camera photographs with technetium Tc 99m pertechnetate give the best resolution. Echo scans can distinguish cystic and solid lesions. A pure cyst is almost never malignant. Lesions that have internal echoes or which are solid on echo scan may be benign or malignant.

Needle biopsy

One of the most helpful procedures in the differential diagnosis of thyroid nodules is needle biopsy. This can be done either by utilizing a thin needle to aspirate material for cytologic examination or a larger cutting needle to take a core of tissue from the nodule. Experience with these procedures indicates that if the cytologic or pathologic appearance is benign, the likelihood of thyroid cancer is extremely small.

Suppressive therapy

Finally, benign lesions may regress after thyroxine therapy, whereas malignant ones are very unlikely to regress. Therefore, marked regression of the nodule following administration of levothyroxine, 0.2 mg orally daily for 3 months or more, is suggestive of a benign lesion.

Decision matrix analysis for separation of benign from malignant lesions

A patient with a thyroid nodule should have a scintiscan. If the lesion is "hot," the likelihood of malignancy is low and the patient should be followed with or without T_4 suppression. These nodules are usually autonomous, and the major concern is the possible development of thyrotixocosis.

If the lesion is "cold," a risk factor evaluation should be done. If the patient is at high risk for cancer—previous therapeutic irradiation, family history of medullary carcinoma, hoarseness, etc—surgery should be done immediately, and no further studies are required. If the patient is at relatively low risk for cancer, fine needle aspiration biopsy should be performed if available. If the results are suspicious or positive, surgery should be performed; if not, the patient should be started on T_4 suppression. If the nodule shrinks, T_4 should be continued indefinitely. If it fails to shrink or if it grows, repeat biopsy or surgery must be considered.

If fine needle aspiration biopsy is not available, echo scan can be done. If the lesion is cystic, it should be aspirated and treated with T_4 suppression. If it is solid or semicystic, surgery should be done. If the lesion recurs several times after aspiration, surgery should be reconsidered.

If this protocol is followed, the incidence of malignant lesions removed at the time of surgery will be in the range of 30-40%. Follicular adenomas cannot be differentiated from follicular carcinomas by either thin needle or cutting needle aspiration biopsy, so that some

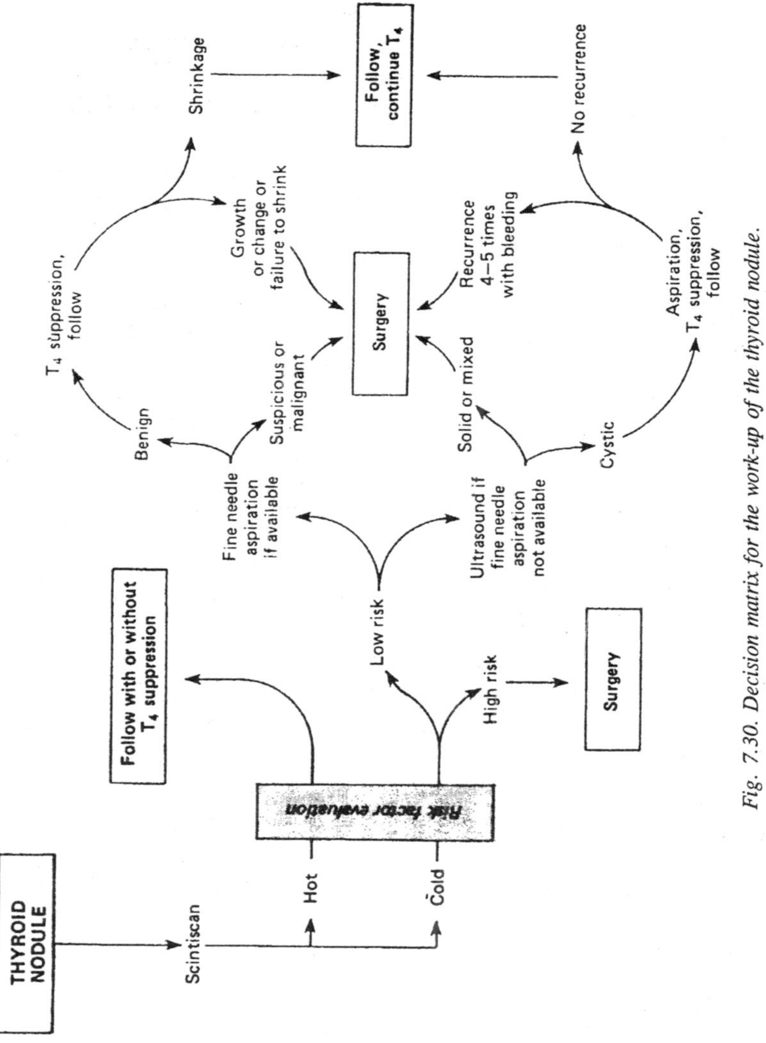

Fig. 7.30. Decision matrix for the work-up of the thyroid nodule.

of these follicular adenomas will have to be removed in order to make certain the patient does not have follicular carcinoma.

Thyroid Cancer

Pathology

Papillary carcinoma

Papillary carcinoma of the thyroid gland usually presents as a firm, solitary, "cold," "solid" nodule clearly different from the rest of the gland. In multinodular goiter, the cancer will be a "dominant nodule"—larger, firmer, and (again) clearly different from the rest of the gland. About 10% of papillary carcinomas, especially in children, present with enlarged cervical nodes, but careful examination will usually reveal a "cold" nodule in the thyroid. Rarely, there will be hemorrhage, necrosis, and cyst formation in the malignant nodule, but on echo scan there will be clearly defined internal echoes that serve to differentiate the semicystic malignant lesion from the nonmalignant "pure cyst." Finally, papillary carcinoma may be found incidentally as a microscopic focus of cancer in the middle of a gland removed for other reasons such as Graves' disease.

Microscopically, the tumor consists of single layers of thyroid cells arranged in vascular stalks, with papillary projections extending into microscopic cystlike spaces. The nuclei of the cells are large and pale and frequently contain clear, glassy intranuclear inclusion bodies. About 40% of papillary carcinomas form laminated calcified spheres, often at the tip of a papillary projection, called "psammoma bodies." These are usually diagnostic of papillary carcinoma. These cancers usually extend by intraglandular metastasis and by local lymph node invasion. They grow very slowly and remain confined to the thyroid gland and local lymph nodes for many years. In older patients, they may become more aggressive and invade locally into muscles and trachea. In later stages, they can spread to the lung. Death is usually due to local disease, with invasion of deep sectors in the neck; less commonly, death may be due to extensive pulmonary metastases. In some older patients, a slowly growing papillary carcinoma will begin to grow rapidly and convert to undifferentiated or anaplastic carcinoma. This "late anaplastic shift" is another cause of death from papillary carcinoma. Many papillary carcinomas secrete thyroglobulin, which can be used as a marker for recurrence or metastasis of the cancer.

Follicular carcinoma

Follicular carcinoma is characterized by the maintenance of small follicles, although colloid formation is poor. Indeed, follicular carcinoma may be indistinguishable from follicular adenoma except by capsular invasion or vascular invasion. The tumor is slightly more aggressive than papillary carcinoma and can spread either by local invasion of lymph nodes or by blood vessel invasion with distant metastases to bone or lung. Microscopically, the cells are cuboidal, with large nuclei arranged around follicles that frequently contain colloid. These tumors often retain the ability to concentrate radioactive iodine, to form thyroglobulin, and, rarely, to synthesize T_3 and T_4. Thus, the rare "functioning thyroid cancer" is almost always a follicular carcinoma. This characteristic makes these tumors more likely to respond to radioactive iodine therapy. In untreated patients, death is due to local extension or to distant bloodstream metastasis with extensive involvement of bone, lungs, and viscera.

A variant of follicular carcinoma is the "Hurthle cell" carcinoma, characterized by large individual cells with pink-staining cytoplasm filled with mitochondria. They behave like follicular cancer. Mixed papillary and follicular carcinomas behave more like papillary carcinoma. Most follicular carcinomas secrete thyroglobulin, and this feature can be used to follow the course of disease.

Medullary carcinoma

Medullary cancer is a disease of the C cells (parafollicular cells) derived from the ultimobranchial body and capable of secreting calcitonin, histaminase, prostaglandins, serotonin, and other peptides. Microscopically, the tumor consists of sheets of cells separated by a pink-staining substance that has characteristics of amyloid. This material stains with Congo red. Amyloid consists of chains of calcitonin laid down in a fibrillary pattern—in contrast to other forms of amyloid, which may have immunoglobulin light chains or other proteins deposited in a fibrillary pattern.

Medullary carcinoma is somewhat more aggressive than papillary or follicular carcinoma but not as aggressive as undifferentiated thyroid cancer. It extends locally into lymph nodes and into surrounding muscle and trachea. It may invade lymphatics and blood vessels and metastasize to lungs and viscera. Calcitonin secreted by the tumor is a marker for medullary carcinoma and can be used both to discover the tumor and as an index of cure or recurrence. About two-thirds of medullary carcinomas are familial, and about one-third are isolated instances of the cancer. The familial tumors may be associated with type II multiple endocrine neoplasia (Sipple's syndrome). In this syndrome, medullary carcinoma is associated with pheochromocytomas and parathyroid adenomas as well as multiple neuromas of the mouth and bowel. At times, only the medullary carcinoma is familial, and no other endocrine glands are involved. The important thing is that if medullary carcinoma is diagnosed, all family members should be screened for possible medullary carcinoma. The best screening test is serum calcitonin after pentagastrin stimulation.

Undifferentiated (anaplastic) carcinomas

Undifferentiated thyroid gland tumors include small cell, giant cell, and spindle cell carcinomas. They usually occur in older patients with a long history of goiter in whom the gland suddenly—over weeks or months—begins to enlarge and produce pressure symptoms, dysphagia, or vocal cord paralysis. Death from massive local extension usually occurs within 6-36 months. These tumors are very resistant to therapy.

Miscellaneous Types

Lymphoma

The only type of rapidly growing thyroid cancer that is responsive to therapy is the lymphoma, which may develop as a part of a generalized lymphoma or may be primary in the thyroid gland. Thyroid lymphoma occasionally develops in a patient with long-standing Hashimoto's thyroiditis and may be difficult to distinguish from chronic thyroiditis. It is characterized by lymphocyte invasion of thyroid follicles and blood vessel walls, which helps to differentiate thyroid lymphoma from chronic thyroiditis. If there is no systemic involvement, the tumor may respond dramatically to radiation.

Metastatic thyroid cancer

Many systemic cancers metastasize to the thyroid gland, including cancers of the breast and kidney, bronchogenic carcinoma, and malignant melanoma. The primary site of involvement is usually obvious. Occasionally, the diagnosis is made by needle biopsy or open biopsy of a rapidly enlarging cold thyroid nodule. The prognosis is that of the primary tumor.

Natural History of Thyroid Cancer

Papillary carcinoma of the thyroid (the most common type) may exist as a microscopic focus in the gland. The incidence of microscopic foci of thyroid cancer at autopsy has varied from 4% in several series in the USA to as high as 25% in persons of Japanese descent living in the Hawaiian Islands. The lesion does not shorten life and is probably of no clinical significance. Woolner has shown that in patients with "occult" or microscopic papillary carcinoma—or small intrathyroidal papillary carcinoma—the survival rate is about the same as that of individuals without thyroid cancer. On the other hand, if there is evidence of extrathyroidal extension of the cancer, life expectancy is markedly affected. Similarly, in persons with small, noninvasive follicular carcinoma, the survival rate does not differ from that of control groups; whereas in the case of follicular carcinoma that has

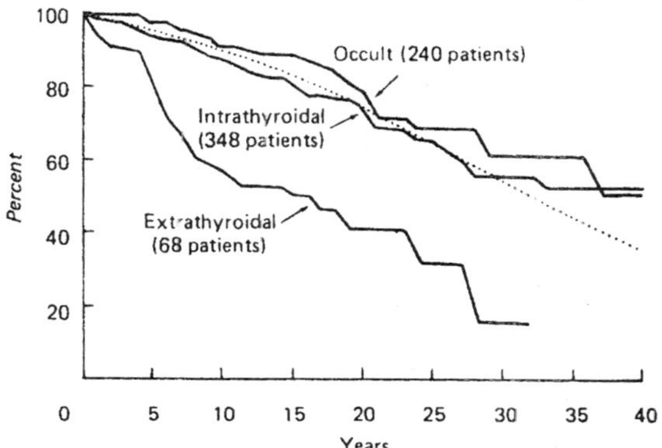

Fig. 7.31. Papillary carcinoma. Survivorship curves for occult, intrathyroidal, and extrathyroidal lesions. Dotted line is curve for normal persons of comparable age and sex.

invaded blood vessels or in which there is evidence of extrathyroidal extension, survival is poor. Medullary carcinoma shows a similar pattern, with good survival for intrathyroidal small lesions and impaired survival for patients with extrathyroidal extension. Patients with anaplastic carcinomas have a very poor prognosis; almost all are dead within 3 years.

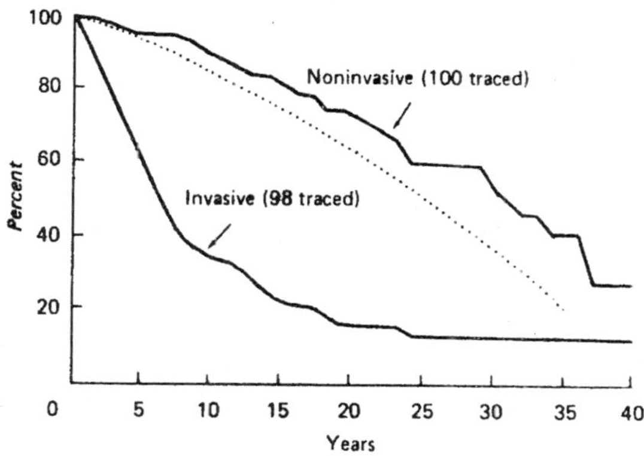

Fig. 7.32. Follicular carcinoma.

Factors Related to Survival of Patients With Thyroid Cancer

Mazzaferri has reviewed factors related to survival of patients with thyroid cancer, including the type of malignancy and the size of the lesion. Papillary, follicular, or medullary carcinomas over 2 cm in diameter have a worse prognosis than small, well-circumscribed lesions. Invasion of the thyroid capsule with local extension into

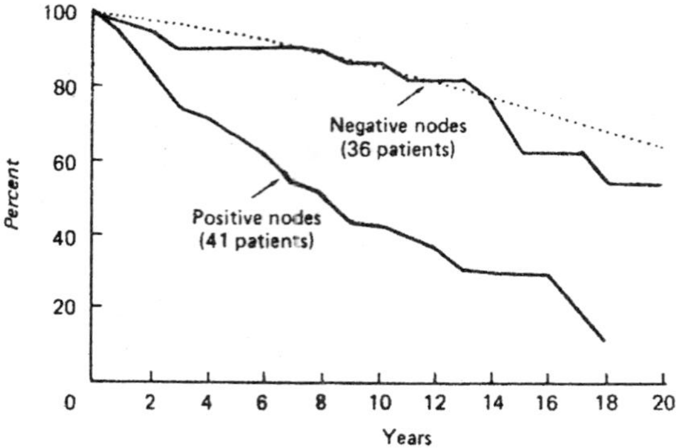

Fig. 7.33. Medullary carcinoma.

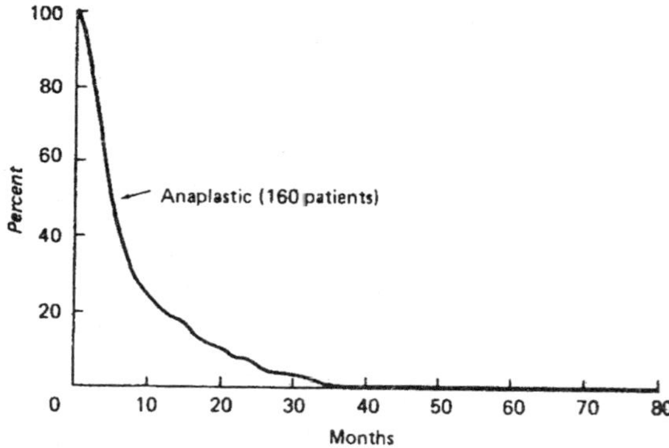

Fig. 7.34. Anaplastic carcinoma.

surrounding muscle and fascia is associated with a poor prognosis. On the other hand, lymph node involvement, particularly in young persons, may not represent a worsening of the prognosis. However, lymph node involvement in patients over 40 years of age implies a worse prognosis. Blood vessel invasion is an adverse prognostic feature. Several studies have shown that papillary carcinoma and follicular carcinoma are more virulent in patients under 7 or over 40 years of age.

Management of Thyroid Cancer

Papillary and follicular cancer

Lobectomy is satisfactory for small (1.5 cm or less) papillary and follicular thyroid carcinomas, but near-total thyroidectomy is required for larger lesions or for cancers with evidence of extrathyroidal extension. Modified neck dissection is indicated if there is evidence of lymph node metastases. Prophylactic neck dissections are not recommended.

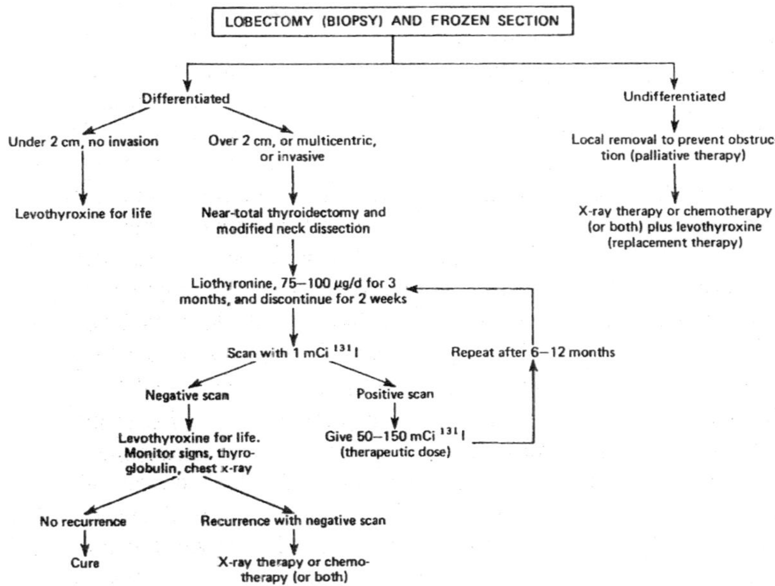

Fig. 7.35. Management of thyroid cancer.

Postoperatively, the patient receives liothyronine (Cytomel), 75-100 μg daily in divided doses for about 3 months; the medication is then stopped for 2 weeks, and the patient is scanned with 1 mCi of ^{131}I. Liothyronine is used for replacement therapy because it is cleared from the blood rapidly: after 2 weeks off therapy, TSH levels are sufficiently high to obtain good scanning studies. If there is evidence of residual radioactive iodine uptake in the neck or elsewhere, radioactive iodine (^{131}I) is effective treatment. The scan is repeated at intervals of 3-12 months until no further uptake is observed; the patient is then maintained on full replacement therapy with levothyroxine, 0.2-0.3 mg daily, to suppress serum TSH to undetectable levels.

Follow-up at intervals of 6-12 months should include careful examination of the neck for recurrent masses. If a lump is noted, needle biopsy examination is indicated to confirm or rule out cancer. Serum thyroglobulin will usually be undetectable after total thyroidectomy and levothyroxine therapy. A rise in serum thyroglobulin suggests recurrence of thyroid cancer. Chest x-ray every 2-3 years may reveal pulmonary metastases, although a rising serum thyroglobulin concentration is a more sensitive indicator. If the, patient does develop a mass in the neck, a rise in serum thyroglobulin, or a mass on chest x-ray, ^{131}I scan should be repeated following the technique outlined above. If the lesions concentrate ^{131}I, this may be useful in treatment; if they do not concentrate ^{131}I, local excision or local x-ray therapy may be useful. For recurrences that cannot be managed by these modalities, chemotherapy should be considered.

Medullary carcinoma

Patients with medullary carcinoma should be followed in a similar way, except that the marker for recurrent medullary cancel is serum calcitonin. Histaminase and other peptides are also secreted by these tumors, but assays for these substances are not generally available. If a patient has a persistently elevated serum calcitonin concentration after total thyroidectomy and regional node dissection, selective venous catheterization and sampling for serum calcitonin may reveal the location of the metastases. If this fails to localize the lesion (as is often the case), the patient must be followed until the metastatic lesion shows itself as a palpable mass or a shadow on the chest x-ray or CT scan. Rarely, metastatic medullary carcinoma can be treated with ^{131}I; usually, however, local excision and levothyroxine therapy are best. Chemotherapy for medullary carcinoma has not been effective.

Anaplastic carcinoma

Anaplastic carcinoma of the thyroid has a very poor prognosis. Treatment consists of isthmusectomy to confirm the diagnosis and to prevent tracheal compression and palliative x-ray therapy. Thyroid lymphomas are quite responsive to x-ray therapy; giant cell, squamous cell, spindle cell, and anaplastic carcinomas are unresponsive. Chemotherapy is not very effective for anaplastic carcinomas. Doxorubicin (Adriamycin), 75 mg/m^2 as a single injection or divided into 3 consecutive daily injections repeated at 3-week intervals, has been useful in some patients with disseminated thyroid cancer unresponsive to surgery, TSH suppression, or radiation therapy. This drug is quite toxic; side-effects include cardiotoxicity, myelosuppression, alopecia, and gastrointestinal symptoms.

X-ray therapy

Local x-ray therapy has been useful in the treatment of solitary metastatic lesions, particularly follicular or papillary tumors, that do not concentrate radioactive iodine. It is particularly effective in isolated nonfunctional bone metastases.

Prognosis

The prognosis is very poor for undifferentiated thyroid cancer but much better for the more differentiated types. Cady et al point out that age and sex of the patient at the time of diagnosis are the most important variables affecting prognosis in differentiated thyroid cancer. They define a "high-risk" group of men over 40 and women over 50 with a recurrence rate of 33% and a death rate of 27%. In contrast, they found that in a "low-risk" group of men under 40 and women under 50, the recurrence rate was 11% and the death rate 4%. These data suggest that although every effort should be made to provide appropriate therapy for all patients with thyroid cancer—surgery, radioiodine, and thyroxine—older patients will not do as well as younger ones despite best efforts.

8

HOMEOSTASIS

Calcium and phosphate are as important as sodium and potassium in the regulation of basic body functions. Inorganic calcium and phosphate are not related closely with respect to most of their essential roles in vertebrate physiology, but these ions are "inseparable" with respect to their involvement in the structure of bone and teeth. Calcium phosphate incorporated into bone matrix serves as the major reservoir for these ions. Consequently a discussion of the endocrine regulation of calcium homeostasis cannot be divorced entirely from the regulation of phosphate. The succeeding discussion of the mammalian pattern of calcium and phosphate homeostasis in mammals will provide a basis for making comparisons to other vertebrates.

IMPORTANCE OF CALCIUM AND PHOSPHATE

Calcium Homeostasis

During embryonic development and in growing mammals calcium is required for formation and growth of bone and teeth, which consist largely of calcium salts, especially calcium phosphate. Calcium ions (Ca^{++}) are essential components of blood plasma, where Ca^{++} functions as a cofactor for certain enzymes involved in the normal blood-clotting process. In addition, many intracellular enzymes require Ca^{++} as a cofactor.

Plasma Ca^{++} levels determine the Ca^{++} concentration of interstitial fluids (usually about 70% of plasma values) and influence the excitability of neurons and muscle cells. The formation of the actinomyosin complex in contracting muscle cells following neural stimulation is initiated by Ca^{++}. The quantity of acetylcholine released from motor end plates is proportional to the concentration of Ca^{++} in the interstitial fluids. Low extracellular Ca^{++} levels, however, may cause stimulation of a motor neuron as a result of the effects of Ca^{++} on cell membrane permeability to other ions such as Na^{+} and K^{+}. Furthermore, low extracellular Ca^{++} levels may stimulate myosin ATPase activity within muscle cells, resulting in uncontrolled contractions.

Recent studies of hormone actions, especially those hormones causing release of other hormones, have shown that the Ca^{++} content of the medium is important to their mechanisms of action. Normal release of hormones and certain other cellular products seems to be dependent upon Ca^{++}, and many neural and endocrine factors that influence hormone secretion may operate through controlling the availability of Ca^{++}.

Calcium regulation

The calcium level in adult mammalian blood plasma is maintained at about 10 mg/dl or approximately 2.5 mM/liter. Somewhat less than half of this calcium is free in the plasma (that is, in ionic form), and the remainder is bound to circulating proteins, primarily albumin. It is this free ionic calcium that is essential to so many important life processes.

Table 8.1. Calcium content of human body

Age (Yr)	Body weight (kg)	Calcium content (g)
1	10.6	100
5	19.1	219
10	33.3	396
15	55.0	806
20	67.0	1078

After birth, calcium is obtained in the diet, absorbed through the small intestine, deposited in bone and teeth or excreted via urine or feces. Urinary excretion of calcium is directly proportional to plasma levels, and little calcium is excreted unless plasma calcium levels exceed normal. Calcium deposited in bone serves as a reservoir to provide adequate plasma Ca^{++} for minute-to-minute regulation of body needs and during acute or chronic periods of dietary deprivation.

Phosphate Homeostasis

Phosphate, like calcium, is an essential component of bone and teeth. Approximately 80% of the total body phosphate is locked into the skeleton as calcium phosphate. In addition, many essential molecules contain phosphate, including structural phospholipids in cellular membranes, nucleic acids, nucleotides and hexose phosphates. Furthermore, phosphate is indispensable for energy storage within cells in the form of ATP or creatine phosphate. Hydrolysis of ATP or guanosine triphosphate (GTP) to form cyclic adenosine 3′, 5′-monophosphate (cAMP) or cyclic guanosine 3′, 5′-monophosphate (cGMP) respectively, is necessary for mediating the actions of many hormones, for neural transmission and for many other cellular processes. The presence or absence of phosphate may determine whether an enzyme has biological activity or not. Finally, phosphate ions play only a minor role as buffers of hydrogen ions in the body fluids, but they are the major buffer system in urine. It is rare that phosphate becomes a limiting factor for an organism, and the usual phosphate disturbances in mammals are a result of excessive levels of phosphate.

Inorganic phosphate (P_i) in mammalian plasma is generally found to be about 3.1 mg/dl (about 1.0 mM/liter). Most of this phosphate (about 80%) is in the form of HPO_4^{-2} with almost 20% occurring as $H_2PO_4^{-1}$ and only a trace as PO_4^{-3}. Henceforth the chemical formula HPO_4^{-2} will be used to represent all of the phosphate ions.

In normal plasma about 90% of the P_i is free ionic (filterable) phosphate, and about 10% is bound to the plasma proteins. In addition to P_i, plasma contains considerable amounts of lipid-bound phosphate and esterified phosphate so that total plasma phosphate is actually about 12.5 mg/dl. Phosphate values vary considerably with diet, age and metabolic state, however, and it is difficult to provide a "normal" value without specifying the conditions under which this "normal" occurs.

Interrelationship of Ca^{++} and HPO_4^{-2}

Calcium and phosphate are regulated in such a way that the product of the plasma concentrations of Ca^{++} and HPO_4^{-2} always equals some constant $[(Ca^{++}) \cdot (HPO_4^{-2}) = k]$. This constant however, may change according to differing physiological states or pathological conditions. For example, k is greater in growing mammals than it is in adults. This relationship between Ca^{++} and HPO_4^{-2} implies that if there is an increase in Ca^{++} a corresponding decrease in HPO_4^{-2} a should follow. Likewise, an increase in HPO_4^{-2} should cause a decrease in Ca^{++}. Although plasma HPO_4^{-2} levels may change without observed alterations in plasma Ca^{++}, this generalization concerning relative amounts of these ions in plasma is useful to illustrate some of the relationships that exist between the regulatory mechanisms governing these ions.

Minute-to-minute adjustments of Ca^{++} and HPO_4^{-2} levels in extracellular fluids are accomplished primarily through a combination of bone destruction (resorption) or formation, absorption of dietary calcium by the small intestine and renal excretion of phosphate.

Bone Formation and Resorption in Mammals

Bone is the major body reservoir for both Ca^{++} and HPO_4^{-2}. Calcium phosphate occurs in the form of small submicroscopic crystals deposited upon an organic matrix composed primarily of collagen fibers. These crystals assume a uniform structural and chemical form known as *hydroxyapatite crystals*. Construction of bone through formation of calcium phosphate is not completely understood, but some of the major features are well accepted.

Bone formation may involve deposition of new hydroxyapatite crystals, *apatite formation*, or simply the additional growth of existing crystals, *mineral accumulation*. Exchange of Ca^{++}, HPO_4^{-2} and water can occur between the surface of these crystals and the extracellular fluids. This exchange is inversely proportional to the size of the crystal. Thus, larger crystals contain considerable amounts of calcium phosphate that cannot engage in free exchange with the extracellular fluids. About 99% of bone calcium phosphate is found in these larger, nonexchangeable stable or "diffusion-locked" crystals. The specific process of bone formation and growth also involves a number of local chemical factors that are responsible for collagen matrix formation, cartilage matrix deposition and formation of hydroxyapatite crystals as well as cellular replication and differentiation.

Cells known as *osteoblasts* (literally, bone forming) are responsible for bone formation. The homologous cell in teeth is termed an odontoblast. The osteoblasts comprise the *endosteal membrane* that lines the cavities within bone, and they synthesize the collagen matrix upon which apatite formation occurs. The factors controlling osteoblast activities are poorly understood. Some osteoblasts give rise to osteocytes that become completely surrounded by bone except for minute channels through which the osteocytes communicate with one another. There seems to be little agreement on the role for osteocytes in calcium-phosphate metabolism, but they may be targets for hormonal regulation.

Resorption of bone may involve either removal of the collagen matrix and/or solubilization of hydroxyapatite crystals with consequent release of Ca^{++} and HPO_4^{-2}. Apparently both processes usually occur during bone resorption. Another bone cell, the *osteoclast* (literally, bone destroying) is primarily responsible for bone resorption. The osteoclast is a large, multinucleate cell and is easy to distinguish from uninucleate osteoblasts. Osteoclasts may arise from either osteoblasts or bone marrow cells, but the details of their formation are not clear.

Endocrine Regulation of Calcium and Phosphate in Mammals

In mammals the regulation of calcium is accomplished primarily by the actions of two hormones: *parathyroid hormone* (PTH), produced by the parathyroid glands, and *calcitonin* (CT), produced by the parafollicular or C cells of the thyroid gland. Parathyroid hormone is a hypercalcemic factor, that is, it causes an elevation in the level of plasma calcium. Release of PTH is increased by low plasma calcium levels and is decreased by elevated plasma calcium levels, but PTH release is not directly influenced by fluctuations in phosphate levels. Calcitonin is a hypocalcemic agent, and its release is directly related to changes in plasma calcium.

A major site of action for PTH is bone, where it may stimulate calcium deposition in bone or bone resorption and release both calcium and phosphate ions into the circulation. Parathyroid hormone also controls phosphate levels by increasing tubular excretion of phosphate by the kidney. Increased calcium reabsorption by the nephron may occur in association with the excretion of phosphate. The net result of bone, renal and possible intestinal actions of PTH is an increase in plasma calcium with a decrease in plasma phosphate levels and a concomitant increase in urinary phosphate. These actions of PTH on bone and possibly on the kidney appear to require the presence of certain derivatives of vitamin D. Dietary calcium uptake through the intestional muscosa is also vitamin D-dependent and may be indirectly influenced by PTH.

Parathyroid Glands

Although the parathyroid glands had been observed previously, Sandstrom rediscovered and named them in 1880. Frequently the parathyroids, which develop like the thyroid from pharyngeal tissues, are embedded within the thyroid glands, for example, in the mouse, cat and man. In other mammals, such as goats and rabbits, they are separate glands located near the thyroid. Some mammals may have as many as four separate parathyroid glands. In addition to the definitive parathyroid glands, accessory parathyroid tissue exists in some species.

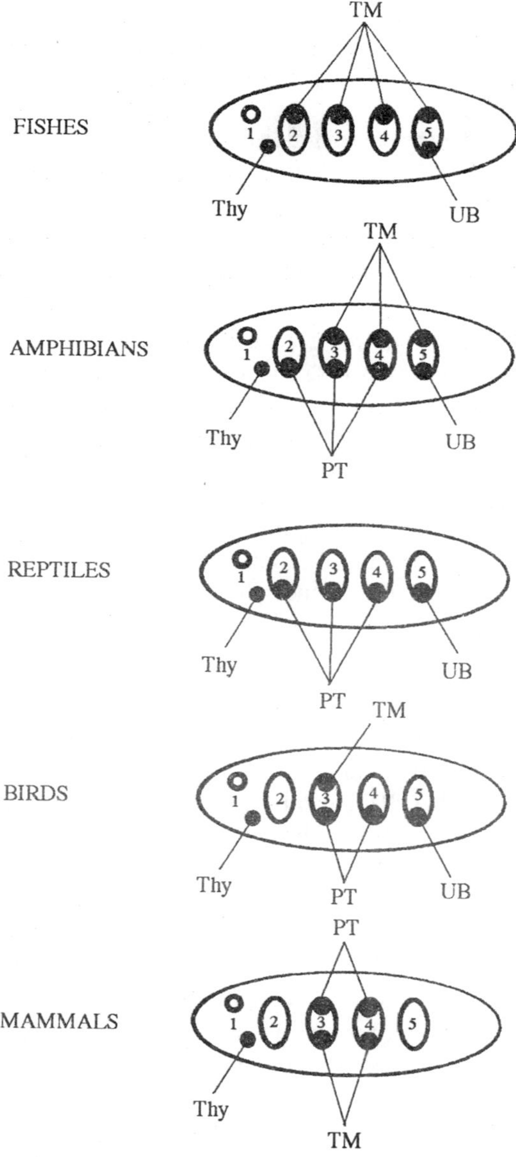

Fig. 8.1. Embryonic origins from pharyngeal pouches for thymus (Tm), thyroid (Thy), parathyroid (PT) and ultimobranchial (UB) tissues in vertebrates. The numbers refer to pharyngeal pouches.

These anatomical peculiarities among different species are reflected in observations that thyroidectomy may result in lowered Ca^{++} plasma levels in some species because of concomitant removal of the parathyroids, whereas parathyroidectomy in another species may not alter plasma Ca^{++} markedly because of the continued presence of accessory parathyroid tissue.

Historically the parathyroid glands have been reported to be derived from pharyngeal endoderm. Parathyroid cells, however, have been reported to arise from neuroectoderm, in the frog *Rana temporaria* suggesting they are also part of the amine precursor uptake and decarboxylation (APUD) series of polypeptide-secreting cells. Indirect evidence also supports a similar origin for PTH-secreting cells in birds and mammals.

Two cellular types have been described in mammalian parathyroid glands: *chief cells* and *oxyphils*. Chief cells are cuboidal with no unique cytological features (other than the presence of granules containing immunoreactive PTH). They are the dominant cellular type and comprise about 99% of the cellular population in most species. In a few species, such as deer, the parathyroids are composed exclusively of chief cells. The oxyphils are rare (sometimes absent) eosinophilic cells rich in mitochondria (hence their name). The function of the oxyphil and the significance of the large number of mitochondria are unknown.

Parathyroid hormone

Parathyroid hormone is a hypercalcemic agent, and its primary role is to increase circulating Ca^{++} and to augment renal phosphate excretion. Mammalian PTH has been isolated and characterized from several species. It is a large polypeptide consisting of 84 amino acids. Apparently all of the biological activity resides in the first 34 amino acids. Parathyroid hormone is synthesized in the chief cells as a larger polypeptide that is later cleaved to release the smaller PTH unit. This synthetic mechanism is essentially like that described for other polypeptide hormones, including growth hormone, prolactin and insulin.

Synthesis of PTH occurs in two steps. A large polypeptide of 115 amino acid residues (pre-proparathyroid hormone [PreProPTH]) is synthesized on the ribosomes of the rough endoplasmic reticulum (RER) but is rapidly cleaved at the amine terminus to a smaller peptide, proparathyroid hormone (ProPTH), of 90 amino acids as it enters the cisterna of the RER. ProPTH travels through the cisterna to the Golgi apparatus where the remaining 6 amine terminal residues of the prohormone are removed. The resulting 84-amino acid sequence of PTH is then packaged into storage vesicles to await release.

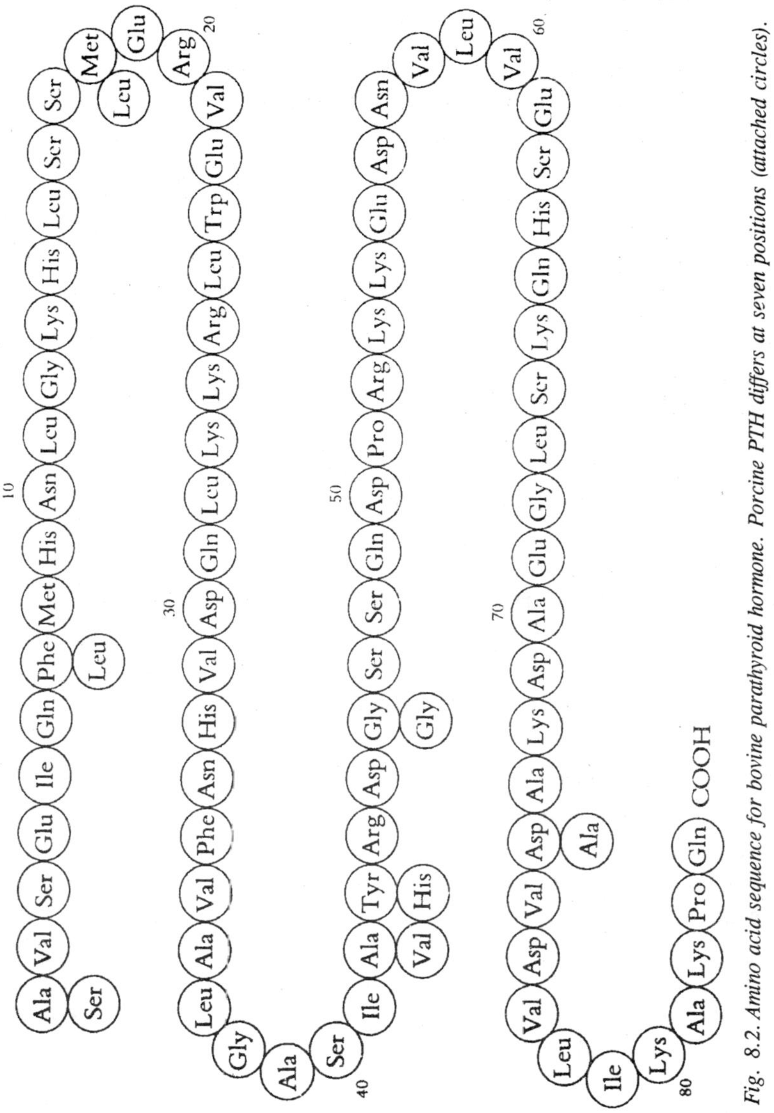

Fig. 8.2. Amino acid sequence for bovine parathyroid hormone. Porcine PTH differs at seven positions (attached circles).

The classical bioassay for PTH is the increase in plasma calcium following administration of parathyroid gland extracts or PTH to normal or parathyroidectomized dogs. More recent bioassays employ parathyroidectomized rats maintained on a calcium-deficient diet.

Parathyroid hormone may occur as peptide fragments in the circulation, varying from 4500 to 9000 daltons. In terms of biological activity, the biological half-life for PTH is about 20 minutes in cattle or rats. The native 84-amino acid PTH peptide is cleared from the blood much more rapidly (2 to 4 minutes) indicating that the remainder of the bioassayable activity is due to persistence of fragments produced by partial hydrolysis of native PTH.

Parathyroidectomy

The effects of removing the parathyroid glands (parathyroidectomy) are variable from species to species. In all cases, parathyroidectomy causes a reduction in plasma Ca^{++} and consequent detrimental muscular effects. As Ca^{++} levels decrease, hyperexcitability of motor neurons and skeletal muscles occurs, resulting in twitches, spasms and, in extreme cases, violent convulsions. This condition is known as *low-calcium-induced tetany*. If prolonged contractions (tetany) of the respiratory muscles occur, death due to asphyxiation may result. The severity of these neuromuscular effects however, differs markedly with respect to species involved and various physiological conditions. For example, exercise raises the body temperature and increases the breathing rate, causing a reduction of blood CO_2. Reduced blood CO_2 in turn alters blood pH (alkalosis) and retards ionization of Ca^{++}. This further reduction in Ca^{++} in a parathyroidectomized animal may precipitate a tetanic seizure. Animals on diets low in Ca^{++} and high in HPO_4^{-2} exhibit tetany following parathyroidectomy more readily than do animals on normal diets.

Parafollicular Cells

The parafollicular (C cells) cells of the mammalian thyroid gland have been identified unequivocally as the source of CT in mammals. However, the actual origin of these cells is unclear. The most generally accepted origin is that proposed by Godwin suggesting that these cells originated from the ultimobranchial body that develops from the sixth pharyngeal pouch (endoderm). These ultimobranchial cells become incorporated into the thyroid gland of mammals as parathyroids frequently do. The ultimobranchial body remains as a distinct separate structure in other vertebrate groups.

The parafollicular cells of the mammalian thyroid exhibit APUD characteristics. Some elegant studies employing cellular chimeras of chicken and quail tissues have confirmed the origin of the CT-secreting cells of the ultimobranchial body from neural crest cells. Parafollicular cells also contain somatostatin and ultrastructurally resemble the D-cells found in the intestinal lining.

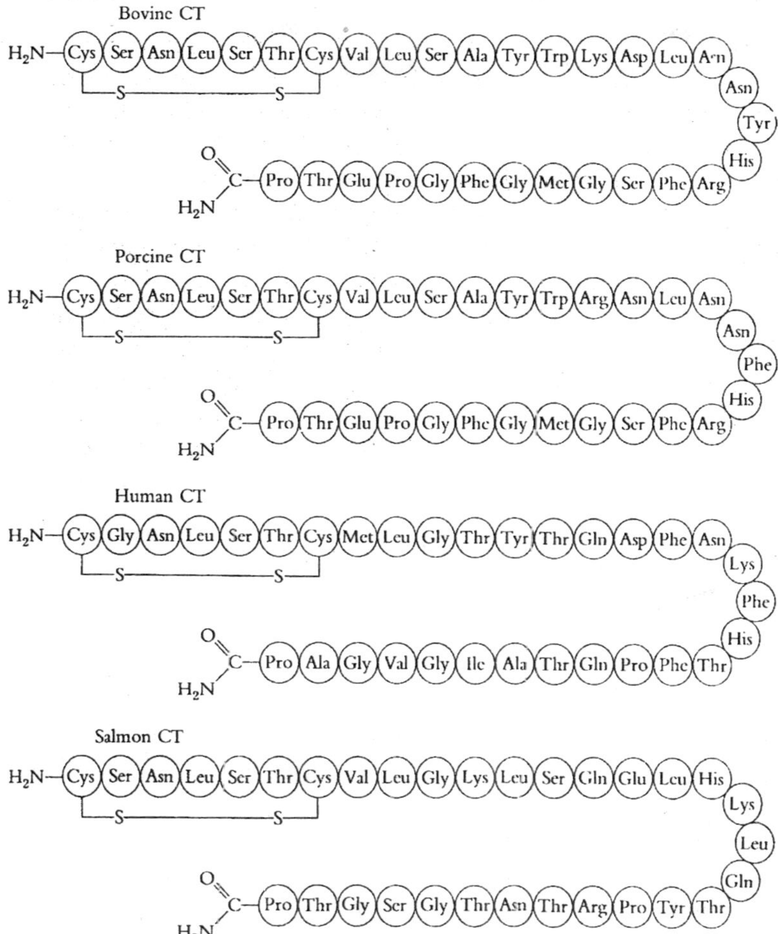

Fig. 8.3. Comparison of amino acid sequences of bovine, porcine, human and salmon calcitonin.

Calcitonin

Production of a potent mammalian hypocalcemic factor by cells of the parathyroid gland was first reported by Copp and associates; it was named calcitonin. Other investigators provided evidence that the source of this hypocalcemic factor was, in fact, the thyroid gland and suggested the alternative name of thyrocalcitonin to reflect its origin. It was soon discovered that ultimobranchial glands of sharks and chickens contained a powerful hypocalcemic factor. These observations were soon reconciled with the ultimobranchial origin of the parafollicular cells of the thyroid that had been shown to be responsible for synthesizing CT.

Mammalian CT is a small, single-chain polypeptide consisting of 32 amino acids with a disulfide link between residues 1 and 7. There are no active fragments of CT, and the entire molecule is necessary for biological activity. It is synthesized as a larger peptide (136 amino acids). The CT sequence actually is cleaved from the middle of the prohormone.

Skin, Liver, Kidney

The actions of PTH on bone and possibly kidney require the presence of a certain derivative of vitamin D. The D vitamins are all steroid molecules that may be obtained through the diet or synthesized by the skin from cholesterol in the presence of adequate sunlight. Generally either vitamin D or vitamin D_3 (cholecalciferol) is converted sequentially by the liver to 25-hydroxycholecalciferol and then by the kidney to *1,25-dihydroxycholecalciferol* (1,25-DHC). This last compound is the necessary cofactor that facilitates the actions of PTH on bone and is itself responsible for intestinal absorption of Ca^{++}. The synthesis of 1,25-DHC by the kidney may be influenced positively by PTH, thus indirectly affecting intestinal absorption of calcium.

Some investigators have proposed that 1,25-DHC should be given the status of a true hormone, a suggestion that, if it were adopted, might require some additional modification of our present concept of "hormone" since at least three tissues (skin, liver and kidney) participate actively in its synthesis.

Interactions of Parathyroid Hormone, Calcitonin and 1,25-Dihydroxycholecalciferol

The major disturbance to calcium homeostasis is the influx of Ca^{++} following ingestion of a meal. In addition, pregnancy and lactation may provide a relatively short-term stress on calcium homeostasis in females. The small intestine, kidney and bone are the primary sites where regulatory hormones (PTH, CT, 1,25-DHC) produce their actions during times of calcium excess or deficiency to maintain calcium homeostasis.

Actions of parathyroid hormone and calcitonin on bone

Two hypotheses have been advanced to explain the actions of PTH on bone. Which of these mechanisms might be the more important has not been determined, and both may be involved in the normal regulatory process. The first hypotheses deals with the proposed action on the

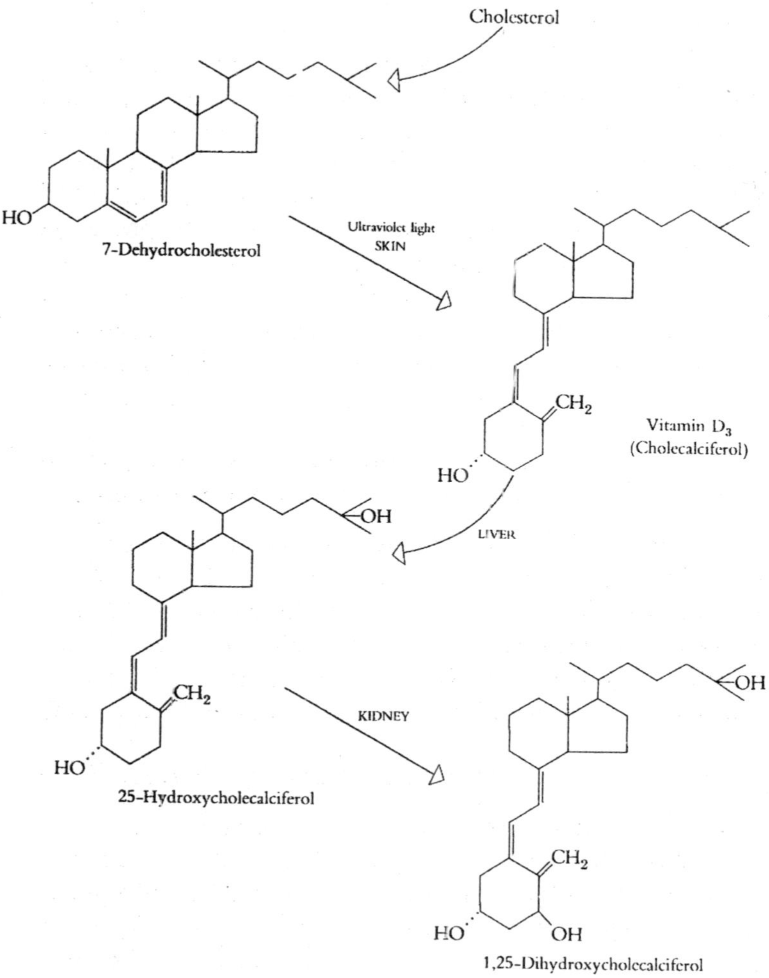

Fig. 6.4. Synthesis of 1,25-dihydroycholecalciferol from cholesterol involving sequential actions of three separate tissues: skin, liver, kidney.

osteoclasts, whereas the second deals with osteoblasts. Osteocytes embedded in the bone matrix may be influenced by both mechanisms and are considered by some researchers to be important targets of PTH.

Osteoclasts

Parathyroid hormone stimulates osteoclast activity and may cause an increase in the number of osteoclasts in bone. Only the mechanism of the former action has been investigated. The action of PTH on the osteoclast first involves intracellular synthesis of small lipid molecules known as prostaglandins. These prostaglandins in turn activate the enzyme adenyl cyclase, resulting in an increase in cAMP within the osteoclast. Apparently an increase in cAMP is related to the release of lysosomal enzymes from cytoplasmic lysosomes. It is the action of these hydrolytic lysosomal enzymes that is responsible for bone resorption and release of Ca^{++} and HPO_4^{-2} to the plasma. Calcitonin apparently interferes with this sequence of events, presumably by interfering with the activation of adenyl cyclase by the prostaglandins.

Osteoblasts

Several investigators have contributed to the following hypothesis suggesting that the osteoblasts of the endosteal membrane are influenced by PTH. According to this hypothesis the osteoblasts contain a pumping mechanism analogous to the one described for intestinal uptake of C^{++}. In this case, however, PTH, through activation of adenyl cyclase and cAMP formation, increases the flux of Ca^{++} into the osteoblast from the bone surface and out of the osteoblast on the blood side of the endosteal membrane. A reverse mechanism has been used to explain the inhibition of bone deposition when PTH stimulates osteoclast activity.

There is evidence to suggest that at normal physiological levels PTH stimulates bone formation rather than bone resorption through an increase in osteoblast activity. The observed effects of PTH on osteoclast activity requires considerable elevation of PTH levels and may occur only under conditions of calcium deprivation. Calcitonin may influence osteoblast functions to a limited extent and alter efflux of Ca^{++} and HPO_4^{-2} from bone. The major site for action of CT, however, appears to be the osteoclast.

Actions of PTH and CT on kidney

Calcium reabsorption by the kidney may be an important physiological action of PTH, although some investigators believe the increased excretion of phosphate is more important. Parathyroid hormone increases the enzymatic formation of 1,25-DHC in kidney and may enhance intestinal calcium absorption indirectly. Estrogens and prolactin also enhance formation of 1,25-DHC, and these actions may be essential in pregnant and lactating mammals, respectively.

Calcitonin antagonizes the action of PTH on bone resorption but does not influence either renal or intestinal processes in normal animals. It has not been possible to demonstrate renal actions of CT except in parathyroidectomized mammals in which CT may cause dramatic changes in phosphate excretion. It will be assumed that the physiological site of action for CT is bone.

Actions of 1,25-dihydroxycholecalciferol on intestinal uptake of Ca^{++}

According to one scheme, Ca^{++} uptake by the intestinal mucosal cell is dependent upon a calcium-binding protein within these cells that is linked to a calcium-activated ATPase. Calcium is actively absorbed by this protein-ATPase complex at the mucosal surface that is in contact with the lumen of the gut. Once in the cell these Ca^{++} ions enter the mitochondria and are somehow transported to the opposite serosal border of the cell where Ca^{++} ions diffuse into the interstitial fluid and then into the blood capillaries. Phosphate ions passively follow the movements of Ca^{++}. Synthesis of a calcium-transport protein (the calcium-binding protein or the calcium-activated ATPase or both) is stimulated by 1,25-DHC.

Actions of calcitonin on intestinal uptake of Ca^{++}

No role has been hypothesized for CT in this intestinal mechanism for calcium uptake. Following the influx of Ca^{++} ions, however, calcitonin is released partly in response to the increase in plasma Ca^{++} but may also be released by *gastrin*, a gastrointestinal hormone released from the gastric mucosa during the early phase of digestion of a meal. This increased level of CT inhibits the action of PTH on bone osteoclasts and favours addition of the absorbed dietary Ca^{++} to the bone matrix.

Calcitonin becomes indispensable during pregnancy and lactation with respect to Ca^{++} mobilization. Its role is apparently to protect the maternal skeleton from excessive destruction in meeting the calcium requirements of fetus or neonate and allow dietary Ca^{++} to be diverted to it. The relative importance of CT in different species varies markedly according to the precise demands for maternal calcium.

A proposed role for parathyroid hormone in acid-base balance

It has been suggested that the parathyroid glands and PTH evolved in terrestrial vertebrates to deal with the problems of acid-base balance when water was in short supply. A change in acid-base regulation was supposedly necessitated by the loss of gills and the ready availability of water in which to dispose of excess hydrogen ions. This hypothesis proposes that calcium regulation was a secondary acquisition and that carbonate and phosphate regulation related to acid-base regulation was primitive. The absence of parathyroid glands and cellular bone among the fishes is believed to support this hypothesis. Although this may seem to be an interesting hypothesis at first, it has been strongly criticized on theoretical grounds. Nevertheless, it would seem that the "sudden" appearance of parathyroids in tetrapods may be related to more than the problem of calcium homeostasis.

MAJOR CLINICAL DISORDERS ASSOCIATED WITH CALCIUM METABOLISM

Hypercalcemia

Excessive plasma calcium levels (greater than 10 mg/dl) can result from a variety of causes. Primary hypercalcemia is characterized by chronically elevated levels of PTH. The most common cause is a single parathyroid adenoma (90% of cases). Carcinomas of the parathyroids are very rare and may account for less than 1% of primary hypercalcemic cases. Approximately one-third of these cases are without overt symptoms, and the remainder exhibit rather generalized and nonspecific symptoms such as weakness, nausea and anorexia. Serum calcium levels are 10.2 to 11.0 mg/dl and are usually accompanied by lowered phosphate levels. Often, elevated serum levels of calcium are not seen because the kidney compensates with increased calcium excretion (hypercalcuria).

Among the many causative factors that can induce secondary hypercalcemia are lung squamous cell carcinoma and renal cell carcinoma that spontaneously secrete PTH. Breast carcinomas may produce vitamin D-like sterols that increase calcium absorption. Chronic immobilization can bring about extensive bone resorption and elevate blood calcium levels.

Hypocalcemia

There are many different conditions that can lead to hypocalcemia. In most cases there is a reduction in PTH secretion. This may be caused by abnormal development of the parathyroid glands, accidental damage or removal by surgery. Hypomagnesemia impairs PTH secretion and indirectly can cause hypocalcemia. In some cases (e.g., psuedo-hypoparathyroidism) the target organs do not respond to PTH and in others an abnormal PTH is secreted that will not activate tissue receptors. Reductions in 1,25-DHC due to failure to convert 25-HC or to synthesize vitamin D also lead to hypocalcemia. Most cases are without serious overt symptoms. Tetany may be inducible under calcium stress but otherwise is absent.

Osteoporosis

This disease is characterized by decalcification of the skeleton resulting in shrinkage, distortion and increased brittleness of the bones. Osteroporosis accompanies aging and is most common among postmenopausal women. Osteoporosis is eight times more common in women than men largely as a consequence of women having smaller bone calcium reserves. Maximum adult bone mass is achieved in women at about age 35 after which calcium losses exceed calcium gain. During menopause there is a gradual reduction in estrogen levels causing a disproportionate decrease in bone mass by allowing increased bone resorption to take place. Estrogen therapy is sometimes employed to arrest this condition. Increased calcium intake by pre- and postmenopausal women may lessen the impact of calcium losses in later life and retard development of osteoporosis.

Paget's Disease

This disorder is caused by increased osteoclast activity resulting in accelerated bone resorption. It occurs in 3% of the population over age 40, and occurs with greatest frequency in people of Western European or Mediterranean descent. About one-third of the people afflicted with Paget's disease do not exhibit any overt symptoms, but the bones become brittle as the disease progresses. Serum levels of calcium and phosphate are usually normal because the excess ions are excreted in the urine. Salmon CT has been used with some success to treat Paget' disease, because it has a much longer biological half-life in mammals than does mammalian CT.

Aspects of Calcium and Phosphate Homeostasis in Nonmammalian Vertebrates

An attempt to discuss the evolutionary aspects of this problem is complicated by major differences, indeed a distinct dichotomy, between the fishes and the tetrapods. Many fishes lack true bone (class Agnatha, class Chondrichthyes), and many of the bony fishes possess acellular bone rather than cellular bone. Scales may provide a source of calcium to bony fishes, and the surrounding waters have been hypothesized to be a more important source of calcium than even the skeleton. Furthermore, parathyroid glands are lacking in fishes as is PTH, and no homologous structures are known. Calcium metabolism appears to be regulated by the pituitary and the corpuscles of Stannius located in the kidneys. The parathyroids and PTH appear suddenly, fully differentiated in the amphibians. Even though most fishes have ultimobranchial glands that contain potent hypocalcemic factors when assayed in tetrapods, these structures do not seem to be important for calcium regulation in fishes. Finally, the comparative approach is hampered by the lack of detailed information concerning calcium and phosphate regulation in fishes. In fact, virtually nothing is known about calcium-phosphate metabolism in cyclostomes.

Estrogenic hormones elevate circulating calcium levels indirectly in females of most nonmammalian species during the process of vitellogenesis associated with oocyte growth. Synthesis of vitellogenic proteins by the liver is stimulated by estrogens. These proteins are released into the blood, through which they travel to the ovaries where they are incorporated into growing oocytes as yolk protein. Vitellogenic proteins readily bind Ca^{++}, indirectly decrease free plasma Ca^{++} levels and consequently stimulate release of more Ca^{++} from reservoirs such as bone. The presence of vitellogenic proteins elevates total plasma levels of Ca^{++}, although free Ca^{++} remains about the same.

Class Agnatha

Cyclostomes do not appear to have any hormonal regulatory mechanisms for calcium and phosphate metabolism. Whatever regulatory mechanisms are employed, they are not as efficient as those found in other vertebrates as evidenced by circulating levels of calcium. Marine species exhibit an intermediate level of plasma Ca^{++} between that of sea water and plasma of tetrapods, whereas the plasma Ca^{++} level in freshwater lampreys is intermediate between that found in plasma of freshwater teleosts and fresh water.

Class Chondrichthyes

These fishes occur in sea water, for the most part, where calcium availability is not a serious problem. Although these fishes lack true bone tissue, calcium salts are added to their cartilaginous skeletons for additional strength, especially in the larger species. The elasmobranch ultimobranchial glands contain a potent hypocalcemic factor when assayed in mammals. Extracts of shark ultimobranchial glands, salmon CT and porcine CT, however, are all ineffective in altering plasma calcium in sharks. Estrogens do not produce any appreciable alteration in plasma calcium as they do in bony fishes and tetrapods. No experimental evidence has been reported for an adenohypophysial factor that would directly affect plasma calcium. Studies of the Nicaraguan freshwater shark might provide some interesting insight since these fishes live in very low calcium environments.

Class Osteichthyes: Actinopterygii

Calcium regulation appears to be under the control of a hypercalcemic adenohypophysial factor and a hypocalcemic factor associated with the corpuscles of Stannius (CS). Hypophysectomy of either *Anguilla* (which has cellular bone) or *Fundulus* (acellular bone) results in a decrease in serum calcium and induction of tetany. Recent studies suggest that the pituitary factor is PRL and that it is the primary hormone controlling calcium levels in low-calcium environments, that is, fresh water. There is also evidence for a parathyroid hormone-like peptide in pituitaries of codfish and eels that has been given the name of *hypercalcin*. This hypercalcemic peptide may prove to be a more potent factor in teleosts than is prolactin, but more studies are needed for confirmation.

The CS are discrete bodies found embedded within the kidney of most actinopterygians. The number of CS varies greatly among the species that have been examined. It has been suggested that the CS are steroidogenic structures, but there is little evidence to support such a claim. Little experimental work has been performed with species other than teleosts, and only a few of these fishes in fact have been examined. Removal of CS (stanniectomy) from the eel *A. anguilla* results in an increase in serum calcium, and administration of CS extracts reduces calcium to normal levels. Surgical stanniectomy of freshwater *A. japonica* is followed by decreased urinary Ca^{++} levels and increased urinary HPO_4^{-2} levels. It is possible that the CS are mainly active in reducing Ca^{++} levels in fish adapted to high calcium environments such as sea water. The active hypocalcemic agent in the CS may be a peptide called *hypocalcin*.

Calcitonin treatment generally has been ineffective in teleosts with respect to lowering calcium, although some positive responses have been reported. The gill may be a major site where CT regulates Ca^{++} transport. Removal of the ultimobranchial glands from *A. japonica* caused effects just opposite to stanniectomy: an increase in Ca^{++} excretion accompanied by a decrease in HPO_4^{-2}, at least suggesting the possible existence of a PTH-like factor.

Salmon CT has been found to be a powerful hypocalcemic agent when tested in avian or mammalian systems, and it is now employed clinically to reduce pathological hypercalcemia (Paget's disease). This potency of salmon CT is related, at least in part, to its persistence in the circulation, that is, its long biological half-life as compared to mammalian CT. It would seem almost that CT evolved prior to the evolution of its function in calcium homeostasis. The correct explanation for this apparent anomaly is more likely related to ignorance of the possible role for CT in fishes and the endocrine regulation of calcium homeostasis.

Some teleost kidneys possess sufficient enzyme necessary for the second hydroxylation of cholecalciferol to form 1,25-DHC. This hormone is apparently not necessary for Ca^{++} uptake by the intestine, but large doses of 1,25-DHC enhances Ca^{++} uptake.

Thyroid state and calcium homeostasis

Serum calcium may be influenced by thyroid state, but the physiological importance of these observations has not been established. Serum calcium levels are decreased in juvenile steelhead trout (*Salmo gairdneri*) that were radiothyroidectomized prior to complete resorption of the yolk sac. It is not possible to distinguish between general effects of radiation that may have damaged some calcium-regulating mechanism and effects due to the absence of thyroid hormones, however. Growth of the skeleton is also abnormal following radiothyroidectomy, supporting an involvement of thyroid hormones at least indirectly. Abnormal skeletal growth could be related to ineffective action of growth hormone in the absence of thyroid hormones.

Class Osteichthyes: Sarcopterygii

The lungfishes lack parathyroid glands and are relatively insensitive to tetrapod calcium-regulating hormones. It is both surprising and somewhat disappointing that in this one case they do not exhibit some tetrapod-like feature. Parathyroid extracts, PTH and salmon CT are all ineffective in altering Ca^{++} levels in the South American lungfish *Lepidosiren paradoxa*. Surprisingly, however, CT and PTH are diuretic and antidiuretic respectively in these fishes. The physiological importance of these observations is not altogether clear, but these substances are certainly not involved in calcium homeostasis. Perhaps CT has some diuretic action in actinopterygian fishes as well.

Class Amphibia

Studies of amphibians have not shed any light on the origin of the parathyroid glands and the evolution of calcium regulation in tetrapods. Ultimobranchial and parathyroid glands are present in the Apoda, Anura and Caudata, and calcium regulation is similar to that observed in other tetrapods. It may well be that the explanation lies within the fossil record of fishes ancestral to the Amphibia or in fossil Amphibia themselves.

Pituitary

A hypercalcemic factor is present in pituitaries of urodele amphibians, and it is of greater importance in the more aquatic species. Parathyroid glands are not present in some urodeles until after metamorphosis, and parathyroids never develop in some permanently neotenic aquatic species such as *Necturus*. These glands are more important in calcium balance of early terrestrial urodeles and especially of anurans.

Ultimobranchial glands

The amphibian ultimobranchial glands develop from the fifth pharyngeal arches. Most urodeles have only one ultimobranchial gland (usually on the left side). In apodans, anurans and some urodeles (for example, *Necturus* and *Amphiuma*) the ultimobranchial glands are paired. Calcitonin has been demonstrated immunologically in the ultimobranchial glands of *Rana temporaria* and *R. pipiens*. Cytologically the ultimobranchial gland consists of one or more simple follicles containing only one epithelial cellular type. This cell has been termed a C cell and is one of the APUD cellular types. The C cell may appear in either a "dark" form (relatively electron dense) or in a less electron-dense "light" form, at least in anurans.

In addition to the C cells, the ultimobranchial gland of *Chthonerpeton indistinctum*, an apodan, contains two types of neuronal endings, presumably cholinergic and purinergic. Although the presence of sympathetic neurons has been demonstrated in the frog *R. pipiens*, most anurans investigated do not exhibit innervation of the ultimobranchial glands. The condition of the urodele ultimobranchial gland with respect to innervation has not been described.

Parathyroid glands

The parathyroid glands appear to develop from the third and fourth pharyngeal pouches as they do in mammals. Careful histological studies conducted on developing *R. temporaria* indicate that the parathyroid cells actually arise from neuroectodermal pharyngeal components rather than from endoderm. In apodan parathyroids, only chief cells are present. Two cellular types have been described for anurans, but they appear to be two forms of chief cells that occur in light and dark phases. In contrast to the condition of apodans and anurans, two distinct cellular types have been reported for parathyroids in the urodele *Cynops pyrrhogaster*. One of these cells is considered to be only a "supportive" cell, however.

Endolymphatic sacs

In addition to cellular bone, the endolymphatic sacs located at the base of the skull appear to be major targets for factors regulating calcium homeostasis. These sacs contain large amounts of calcium carbonate and may be important reservoirs of these ions, particularly during metamorphosis and subsequent ossification of bones. These structures may also provide bicarbonate ions for buffering the blood following the dissociation of calcium carbonate.

Calcium and phosphate homeostasis

The parathyroids, pituitary gland and ultimobranchial glands have been implicated in calcium and phosphate metabolism. The vitamin D complex seems to be related to at least some actions of PTH in amphibians. Removal of the ultimobranchial glands in frogs generally causes an increase in osteoclast activity and a consequent increase in blood calcium levels. The ultimobranchial glands also prevent removal of calcium from the endolymphatic sacs and block uptake of calcium through the gut.

Parathyroidectomy causes a decrease in plasma Ca^{++} in anurans and in the newt *C. pyrrhogaster*. No changes in serum Ca^{++} occur following parathyroidectomy of immature giant salamanders, *Megalobatrachus davidianus* and additional studies are needed in urodeles to clarify this apparent dichotomy. Administration of bovine PTH to frogs increases plasma Ca^{++} and decreases plasma HPO_4^{-2}, implying that the kidney may also be a target for PTH. There appears to be a seasonal variation in the activity of parathyroid glands in frogs that also influences the activity of the ultimobranchial bodies. In general the amphibians regulate plasma Ca^{++} and HPO_4^{-2} similarly to mammals.

Class Reptilia

The regulation of calcium and phosphate in lizards has been well studied, and in recent years these investigations have been extended to include snakes and turtles. Both ultimobranchial glands and parathyroids have been described, and regulation of calcium is essentially like that in other tetrapods. The major sites for endocrine regulation of calcium metabolism are the kidney, cellular bone and the endolymphatic sacs, which in lizards, as in amphibians, are important reserves of calcium and carbonate ions.

Ultimobranchial glands

The ultimobranchial glands of reptiles are located near the thyroid and parathyroid glands. They are small glands consisting primarily of follicles. All the reptilian ultimobranchials are innervated, although the nature of these neuronal endings has not been examined extensively. In crocodilians, chelonians and in some snakes the ultimobranchial glands are paired. In lizards only the left gland persists. Seasonal changes in the cytology of ultimobranchial glands have been reported for a few species, but the relationship between these changes and physiological and environmental parameters has not been ascertained.

Parathyroid glands

Four reptilian parathyroids develop from the third and fourth pharyngeal pouches as described for the amphibians, and they resemble the mammalian parathyroids cytologically. Adult lizards and crocodilians have only one pair of parathyroid glands, whereas snakes and turtles retain all four glands. In addition to cellular cords, the presence of follicles containing PAS (+) material is a common feature of reptilian parathyroids. The functional significance of these parathyroid follicles is not known.

Calcium phosphate homeostasis

Parathyroidectomy of lizards and snakes causes a marked decrease in plasma Ca^{++} accompanied by tetany. However, there is little or no change in circulating Ca^{++} in turtles following a similar operation, and tetany does not occur in turtles. This insensitivity of turtles to parathyroidectomy is apparently a consequence of the immense calcium reservoir represented by the shell. Treatment with mammalian PTH causes increased plasma Ca^{++} and urinary HPO_4^{-2} as well as decreased urinary Ca^{++} and plasma HPO_4^{-2} in both lizards and turtles.

The renal action of mammalian parathyroid extracts on HPO_4^{-2} excretion is marked in parathyroidectomized snakes (four species of *Natrix*) although calcium excretion is probably not affected significantly. Thus it appears that reptiles possess basically the same regulatory control of calcium homeostasis as exhibited by mammalian parathyroids.

The role of CT in reptiles is uncertain. Extracts of reptilian ultimobranchial glands produce hypocalcemia when injected into rats, and salmon CT has been shown to lower Ca^{++} in the green iguana. As with fishes and amphibians, when reptilian ultimobranchial factors are tested in mammals the presence of hypocalcemic factors is noted, but their endogenous physiological roles are not known.

Class Aves

There are typically four parathyroid glands in birds, and ultimobranchial glands are present. Cytologically the parathyroid glands resemble those of mammals, and the ultimobranchial cells are like the mammalian C cell. The regulation of calcium and phosphate homeostasis is typically mammalian with only minor differences. It should be noted, however, that few avian species have been investigated thoroughly, and most of the studies have been conducted with domestic birds.

Parathyroid glands

Avian parathyroid glands are separate and usually distinct, except for a few species, such as the domestic chicken, in which some fusion of the separate parathyroids may occur. Cytologically the parathyroid glands contain only chief cells; no oxyphils have been reported.

Ultimobranchial glands

The ultimobranchial glands are usually separate structures although some fusion with the thyroid gland occurs in the pigeon. There are both light and dark cells in chicken ultimobranchials. The light cell is more abundant and resembles the mammalian C cell cytologically. A rich vagal innervation (parasympathetic) has been described for the chicken, but its importance has not been elucidated.

Calcium and phosphate homeostasis

Parathyroidectomy in birds usually causes marked hypocalcemia accompanied by tetanic seizures and death within 24 hours. This degree of sensitivity to parathyroidectomy is not manifest on this time scale in amphibians or reptiles and may be related to the much higher body temperature of birds. Treatment with mammalian parathyroid extracts produces marked increases in plasma Ca^{++}. Dietary deprivation of Ca^{++} or vitamin D causes marked hypertrophy and hyperplasia of the parathyroid glands, whereas high Ca^{++} diets result in regression of the parathyroids. As in mammals, estrogens and parathyroid extracts both increase levels of 1,25-DHC in Japanese quail.

Mammalian parathyroid extract stimulates renal excretion of phosphate in normal starlings but does not seem to alter renal treatment of calcium. The effect of mammalian parathyroid extract or PTH on avian bone is similar to that described for mammals.

Calcitonins that are structurally similar to mammalian CT have been isolated from chickens and turkeys and are potent hypocalcemic agents in both birds and mammals. Dietary levels of calcium are directly related to ultimobranchial activities, and high levels of Ca^{++} cause marked stimulatory changes.

9

PANCREAS

Primarily because of the lack of adequate microanalytical methods, our knowledge of the development of endocrine function in the pancreas is grossly deficient. However, there have recently been developed chromatographic and immunological methods for the detection and measurement of microquantities of insulin. We may hope that these methods will soon be adapted to embryological materials.

No direct analyses of islet function in embryological material have been made other than the studies of Banting and Best (1922) and Fisher and Scott (1934) demonstrating the presence of relatively large amounts of insulin in the pancreas of the five-month-old fetal calf. Therefore, the following discussion will center around an evaluation of the available indirect evidence bearing upon the subject of insulin elaboration and secretion by embryonic islet tissue. Two main classes of evidence will be considered: (1) that derived from considerations of the physiology of the embryo and mother and (2) histological and histochemical evidence.

PHYSIOLOGICAL EVIDENCE

Inference from Carbohydrate Metabolism

In view of the key role of insulin in regulating carbohydrate metabolism, we are tempted to correlate our knowledge of the morphological maturation of the islets of Langerhans with supposed physiological consequences of insulin secretion in the embryo and thus to reach certain conclusions concerning functional activity by the beta-cells. However, in view of the great complexity of the mechanism regulating carbohydrate metabolism, the resolution of such an analysis with respect to the specific components of the system (e.g., islets of Langerhans) is far from critical.

Aron and his co-workers early assumed the working hypothesis that there is a direct and a causal relationship between insulin secretion and liver glycogenesis. They demonstrated an almost infallible correlation between the onset of, or a rapid upsurge in, liver glycogen deposition and the morphological maturation of the islets of Langerhans in a variety of vertebrate species.

More recently a consistent relationship between liver glycogen deposition and beta-cell maturation has not been demonstrated. Corey (1935) found that glycogen begins to accumulate in the fetal rat liver about the sixteenth day of gestation. This precedes histological and histochemical differentiation by two days and would seem to rule out a causal relationship between the two phenomena. Dumm (1943), however, has reported a rapid increase in glycogen in the liver of the fetal rat beginning on the nineteenth day of gestation and continuing until birth. Liver glycogen begins to increase rapidly in the 25-day rabbit fetus, but mature beta-cells are not identified until a week after birth. In the fetal guinea pig, on the other hand, the first appearance of liver glycogen (on the forty-fifth day of gestation) follows closely upon the initiation of islet maturation.

In mammals, therefore, the situation is complex and inconsistent and suffers badly from a lack of studies on fetal physiology. Interpretation of the data is rendered more precarious in view of the current controversy concerning the exact relationship between insulin and liver glycogenesis.

The situation is somewhat more suggestive in the chick. Lee (1951) followed the glycogen content of the liver of the chick embryo between the seventh and eighteenth days of incubation and concluded that there was no correlation between the results obtained and the supposed appearance of insulin in the beta-cells on the twelfth day of incubation. Examination of Lee's data, however, suggests a sharp and rather continuous increase in liver glycogen, beginning on the thirteenth day of incubation, and continuing (with slight interruption on day 17) until the eighteenth day, when a sudden sharp fall occurs. Leibson (1950) and Konigsberg (1954) found an increase in liver glycogen between the twelfth and fourteenth days of incubation and suggested that this is associated with differentiation of the beta-cells of the embryo.

This interpretation is supported by the finding that the chick embryo has the capacity to regulate its blood sugar prior to hatching,

Hanan (1925) reported that the chick embryo of 14-16 days of incubation is able to overcome the hypoglycemic and hyperglycemic effects of insulin and glucose respectively, the blood sugar level returning to normal within 4 hours after treatment. More recently Zwilling (1948, 1951) has found that the 12-day-old chick embryo is able to overcome hypoglycemia (and other anomalies of carbohydrate distribution) induced by the injection of insulin into the yolk sac prior to this time. Likewise, Konigsberg (1954) found the twelfth day to be critical in the regulation of the blood sugar of the hypophysectomized embryo.

The twelfth day appears, then, to be critical in the carbohydrate metabolism of the chick embryo. Caution must be exercised, however, in attributing this change to the functional differentiation of the beta-cells. Indeed, the total evidence suggests that several factors affecting carbohydrate metabolism, including insulin secretion, probably became active at this time. The situation has been further complicated by the recent report of Lieure (1957) that mature beta-cells first appear on the seventeenth day of incubation, rather than on the twelfth day as indicated above.

Diabetes and Pregnancy

Studies on the course of maternal diabetes during pregnancy have caused much speculation concerning the functional activity of the fetal islets of Langerhans. The argument centers around the hypothesis that the fetal pancreas may produce insulin which is passed, via the placental circulation, to the mother where it exerts a hypoglycemic effect. Studies made by Carlson and his co-workers showed that dogs depancreatized during late pregnancy (within three weeks of term) failed to become diabetic until after the fetuses were born, aborted, or otherwise separated from the maternal animal. These results have been adequately confirmed and shown to be correlated with the histological maturation of the islets of Langerhans in the fetal pancreas. They are generally interpreted to indicate either the passage of insulin (or other hormonal factors) from the fetuses to the mother, or the utilization of excess carbohydrate by the developing fetuses. This latter possibility remains a most serious objection to these early studies attempting to demonstrate functional activity in the fetal islets of Langerhans on the basis of an alleviation of maternal diabetes.

Experimental studies on forms other than the dog have failed to demonstrate a consistent or appreciable effect of pregnancy upon maternal diabetes. In the rat, for example, maternal hyperglycemia continues unaltered throughout pregnancy until just before term when a slight reduction is noted. Since the rat placenta is permeable to insulin by the sixteenth day of gestation, and, since the rat fetus is capable of insulin secretion at least by the 18th day of gestation, we are led to conclude that the absence of a prolonged reduction in the severity of maternal diabetes is not a valid reason to exclude the possibility of fetal insulin secretion. In fact, the fetal effect probably depends upon several factors such as the degree of placental permeability to insulin in the fetal-maternal direction, the amount of insulin elaborated and secreted by the fetuses, and the time of onset of maternal diabetes. That this latter factor is, in fact, of great importance is indicated by our finding that the beta-cells of fetal rats developing in a diabetic maternal system are completely depleted of insulin shortly after the beta-cells become functional (on the eighteenth day of gestation) and remain in a state of functional exhaustion during the remainder of the gestational period. Under such conditions the fetal islets would scarcely be expected to have an appreciable effect upon maternal glycemia and, in fact, under these conditions the fetal glycemia is a function of the dominating maternal hyperglycemia. Britton (1930), however, has demonstrated that the fetal kitten has some capacity to compensate for fluctuations induced in the maternal blood sugar level during late pregnancy. This self-regulative capacity suggests that insulin secretion by the fetus may occur during late pregnancy in the cat.

The glucose tolerance test is a more sensitive indicator of possible fetal islet function in pregnant animals. On the hypothesis that pregnancy might be expected to shift the glucose tolerance of a depancreatized animal back toward normal at such a stage as the fetuses become capable of insulin secretion, the following experiment was performed: Glucose tolerance was followed in both intact and depancreatized nonpregnant rats and in pregnant rats depancreatized between the twelfth and twenty-first days of gestation. The results are summarized in fig. 10.1A and indicate that pregnancy results in a rather sharp (statistically significant) shift of the glucose tolerance of a depancreatized maternal animal toward normal beginning on the eighteenth day of gestation. The possibility of an effect due to fetal utilization of the sugar is eliminated (1) by the administration of large doses of glucose (500 mg/200 gm) on a body weight basis (such that increased fetal size and carbohydrate utilization are automatically compensated for) and (2) by the use of D-xylose instead of glucose in a similar series of experiments. We interpret these results to indicate the initiation of insulin secretion

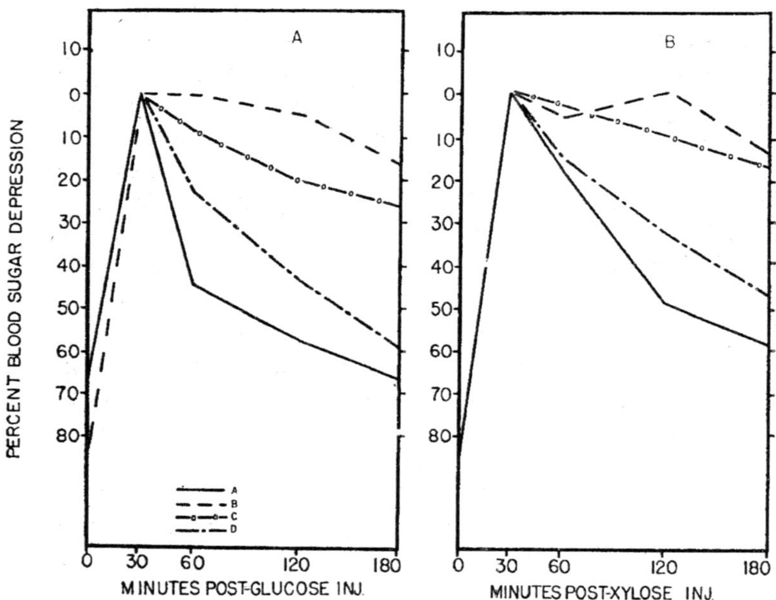

Fig. 9.1. Graph A: Curves showing the shift of glucose tolerance of depancreatized rats toward normal between the 18th and 21st days of gestation. Graph B: Xylose tolerance curves. Legand: A, non-pregnant intact rats; B, non-pregnant depancreatized rats; C, depancreatized rats pregnant 17 days or less; D, depancreatized rats pregnant 18 days or more.

(in "significant" amounts, at least) on the eighteenth day of gestation. This conclusion is in agreement with the histological and histochemical findings discussed below.

Histological and Histochemical Evidence

Histological description of the developing islets under normal and experimental conditions remains the most direct and revealing approach to the problem of islet cell function. However, such studies suffer from a lack of assurance that histological differentiation is a true and critical criterion of functional activity, even though the beta- cells show a certain specificity for the stains used. Recently Barrnett and co-workers (1955) have adapted a method for the demonstration of protein bound sulfhydryl and disulfide groups to the histochemical demonstration of insulin and have shown that, under proper conditions of fixation, etc., the staining reaction of the beta-cells depends upon their insulin content. We have used this method on embryological material in two cases and have found it to be valuable in supplementing routine histological procedures.

Experimental modification of the islet cell histology by variation of the blood sugar level has given somewhat more critical insight into the problem of the secretory capacity of the fetal beta-cells. In adult animals, an abundance of evidence has established fairly conclusively that the primary stimulus for the release of insulin from the beta-cells is the blood sugar level itself, insulin being secreted in response to a high blood sugar concentration. Moreover, prolonged or extreme stimulation of the beta-cells, as by continuous injection of glucose, results in progressive changes indicative of functional strain, including degranulation and vacuolization. In extreme cases, cellular necrosis and experimental diabetes can be produced. In view of this specific and dramatic response of the functional beta-cells to the blood sugar level, it is reasonable to assume that whenever the embryonic beta-cells become functionally mature they will respond to the specific secretory stimulus as do the beta-cells of the adult. Such a study has been carried out with rat fetuses developing under the stress of a diabetic maternal system (alloxan diabetic rats were allowed to become pregnant; maternal blood sugar levels during pregnancy ranged between 195.6 and 464.6 mg %. Fetal glycemia corresponded closely to that of the mother). Under these conditions the beta-cells of the fetal islets begin to show evidence of degranulation and vacuolization on the eighteenth day of gestation. This corresponds closely to the time of initiation of insulin elaboration, as indicated by histochemical procedures. These changes become more pronounced and are paralleled by a loss of histochemically demonstrable insulin from the beta-cells. By the twenty-first day of gestation extreme vacuolization of the cells is noted and the beta-cells are almost devoid of histochemically demonstrable insulin as compared to the controls.

Thus, we are led to conclude that the capacity to secrete insulin matures on the eighteenth day of gestation in conjunction with the initiation of insulin elaboration. This does not necessarily mean that the beta-cells normally begin to secrete insulin at this time. Rather, the histological and histochemical picture is such as to suggest that the beta-cells are engaged primarily in insulin elaboration prior to birth. If the fetus requires insulin m its relatively homeostatic environment, sufficient quantities may be provided by the maternal animal, although the possibility of some secretory activity by the fetal islets cannot be ruled out. Fetuses developing in a hypoglycemic environment (of insulin-treated mothers) show a greater accumulation of stainable insulin than normal embryos, suggesting an even greater inhibition of secretion under these conditions, and thus implying some secretion by the normal embryo.

A descriptive study of the development of the islets of Langerhans in the salamander, *Ambystoma opacum*, reveals that histological maturation of the beta-cells does not occur until late larval life, or just before metamorphosis. Significantly, histochemically demonstrable insulin appears at the same time as histological maturation. These results indicate that, in some species, at least, a major portion of the larval life can be passed before the islet tissue becomes functional.

10

Gastrointestinal Hormones

With the identification by Bayliss and Starling of the first blood-borne chemical messenger, secretin, produced by the duodenal mucosa, and the proposal for *gastrin*, from the antral stomach a few years later by Edkins, the existence of endocrine regulation of the mammalian digestive system was established and the discipline of endocrinology was born. Ironically the gastrointestinal (GI) hormones have been among the last to be chemically characterized, and it was only after isolation of the purified hormones in the 1960s that it became possible to confirm the diffusely distributed cellular types responsible for their secretion. Because of the lack of precision in observations, the complications of sorting out neutral and proposed endocrine factors, effects of known pharmacological agents and rapid proliferation of unsubstantiated factors, the entire area of GI endocrine research was easily over-shadowed when the nature of certain medically important hormones was discovered. Research with corticosteroids, reproductive hormones, thyroid hormones, insulin and epinephrine occupied the mainstream of endocrine research, whereas advances associated with GI factors were uncommon.

In recent years, there has been rapid expansion in GI peptide research, and there are now many peptides that have been isolated and chemically characterized. Yet we do not understand the physiological roles for many of these peptides. For others, the assignment of their physiological roles changes so frequently that it is bewildering. Even their names seem to be in a state of flux. Some GI peptides appear to function as classic hormones, others are paracrine secretions and still others may be neurotransmitters. Many GI peptides have been identified within the central nervous system and in certain endocrine glands. For simplicity, all of these regulatory substances are referred to here as GI peptides.

Endocrine regulation of digestion must be considered an integral portion of the entire digestive process. Consequently it is necessary to review the entire digestive process to illustrate the regulatory actions of GI peptides. The following discussion emphasizes the human digestive system, including the mouth, pharynx, esophagus, stomach and small intestine. In addition to the roles played by these portions of the alimentary canal, digestion is aided by three essential exocrine glands: (1) the several pairs of salivary glands that secrete into the mouth, (2) the liver and (3) the exocrine pancreas; the last two secrete materials into the small intestine. The reader should keep in mind the many anatomical and functional differences that exist between the human digestive system and those of mammalian carnivores and herbivores with respect to the specific details of the following account.

Human Digestive System

The first detailed and systematic knowledge of human digestive processes came from the observations of William Beaumont on his patient Alexis St. Martin, a French Canadian who, while visiting Fort Mackinac, Michigan, was accidentally shot in the chest from close range by a shotgun; two ribs were fractured, the lungs lacerated and the stomach perforated. Although Beaumont assumed that St. Martin would not live the night, he miraculously survived. The wound in St. Martin's stomach, however, never healed completely, resulting in a permanent opening to the outside (a *gastric fistula*) through which Beaumont was able to observe the progression of gastric digestion under varying conditions over a period of years. It was these pioneering observations by Beaumont that stimulated much of the later interest in gastric physiology.

The accidental production of a gastric fistula in St. Martin provided the inspiration for a variety of surgical techniques, including production of gastric fistulas and gastric or intestinal pouches (isolated pouches no longer connected with the lumen of the gut). Transplantation of denervated pouches or pieces of digestive tract or pancreas to sites under the skin where revascularization can occur has enabled investigators to separate endocrine and nervous regulatory mechanisms. Finally the development of crossed circulatory systems between experimental animals was used to confirm the transfer of chemical factors (hormones) through the blood to target tissues. *In-vitro* studies

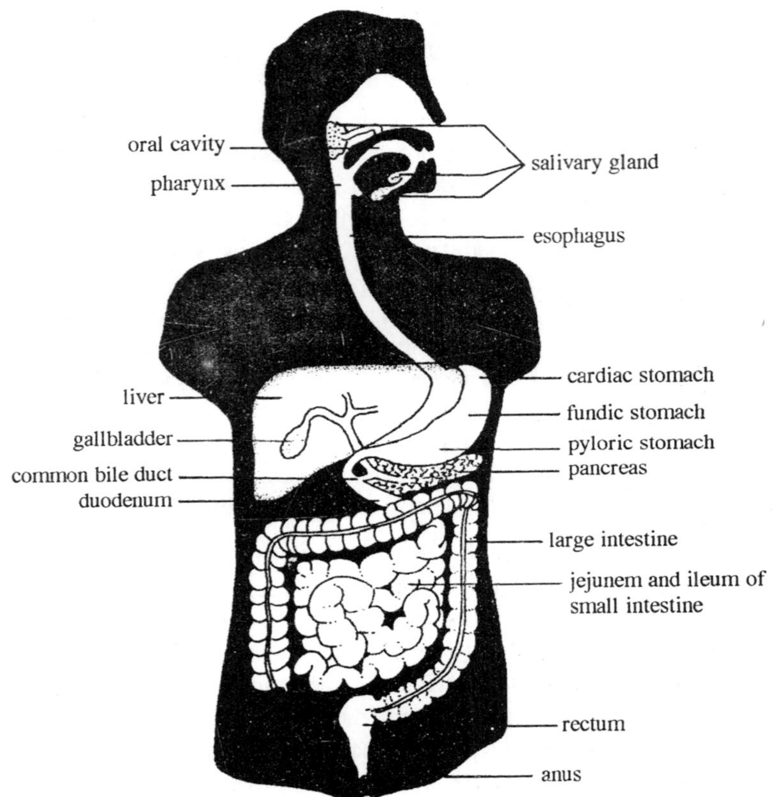

Fig. 10.1. The human digestive system.

of pancreatic slices or mucosal tissues from various regions of the gut have been employed profitably to ascertain details of the actions of the various GI peptides and factors regulating their release.

Oral Events

When food is ingested it is first torn and ground by the teeth and is mixed with *saliva* secreted from the exocrine salivary glands. Salivation is under direct neural control and can be stimulated by sight, smell or simply the thought of food or by the presence of food in the mouth. Saliva is a basic fluid containing the hydrolytic enzyme *salivary amylase*, which hydrolyses starches to disaccharides. The saliva and partially digested chewed food are mixed thoroughly to form a bolus that is pushed back into the pharynx by the tongue. After entry into the pharynx the bolus is swallowed by a reflexive series of events that propel it down the esophagus and into the stomach.

Gastric Events

Upon entering the stomach the bolus of food is in an acidic environment due to secretion of *hydrochloric acid* by glands located in the *mucosa* (epithelial lining) of the *fundus* or main body of the stomach. The specific source of HCl is the *parietal cell* of the mucosa. These cells can be distinguished by staining fixed tissue sections with modification of the Zimmermann technique, following which the parietal cells appear yellow. Hydrochloric acid reduces the pH of the stomach contents, thereby activating the proteolytic enzyme *pepsin*, which is secreted in an inactive form, *pepsinogen*, by the *chief cells* of the gastric glands. Chief cells are stained blue by the modified Zimmermann technique. The acidity of the gastric contents softens fibrous material, kills some bacteria ingested with the food and extracts calcium salts from bone and cartilage. In addition, HCl brings about slow inactivation of salivary amylase as the acid penetrates the bolus and blocks further starch digestion.

It is not known how the parietal cell is able to secrete concentrated hydrogen ions (HCl) without damage to itself. The presence of tight junctions between the mucosal cells and the high mitotic rate observed for gastric mucosal cells in part explain how the stomach contains such a caustic solution without causing irreparable destruction. These tight junctions prevent the penetration of acid and proteolytic enzyme (pepsin) into the mucosa. This outer layer is continuously replaced by mitotic activity to maintain the integrity of the mucosal barrier.

Gastrin theory and acid secretion

Edkins in 1905 showed that extracts prepared from the most posterior portion of the stomach, the *antrum*, stimulated acid secretion by the fundic glands, and he suggested the name of *gastrin* for the active substance in these extracts. He found no gastrin activity in extracts prepared from the fundic portion of the stomach. Edkin's gastrin hypothesis temporarily lost credibility with the discovery of *histamine*, a potent stimulator of gastric acid secretion, and the demonstration of histamine in extracts of the gastric mucosa. It was almost 30 years before it was shown that histamine-free extracts from the mucosa of the antral portion of the stomach possessed the ability to stimulate acid secretion by the parietal cells. Nevertheless it was not until gastrin was finally isolated and characterized chemically that the term gastrin theory was discarded.

At least two similar peptides with gastrin activity have been isolated (gastrin I and gastrin II) from the antral stomach. Both of the molecules are composed of 17 amino acids; their only difference is that the C-terminal tyrosine of gastrin II is sulfated. A molecule of 34 amino acids has also been found in the circulation; it consists of gastrin I or II together with a third 17 amino acid peptide component. This big gastrin constitutes only about 5% of the circulating gastrin. The preprohormone for gastrin is composed of 104 amino acids. This

	Secretin	Glucagon	GIP	PZCCK	Caerulein	Human Gastrin II
1	His	His	Tyr			
2	Ser	Ser	Ala			
3	Asp	Gln	Glu			
4	Gly	Gly	Gly			
5	Thr	Thr	Thr			
6	Phe	Phe	Phe			
7	Thr	Thr	Ile			
8	Ser	Ser	Ser			
9	Glu	Asp	Asp			
10	Leu	Tyr	Trp			
11	Ser	Ser	Ser			
12	Arg	Lys	Ile			
13	Leu	Trp	Ala			
14	Arg	Leu	Met			
15	Asp	Asp	Asp			
16	Ser	Ser	Lys			
17	Ala	Arg	Ile			(1) Glu
18	Arg	Arg	Arg			(2) Glu
19	Leu	Ala	Gln			(3) Pro
20	Gln	Gln	Gln			(4) Try
21	Arg	Asp	Asp			(5) Leu
22	Leu	Phe	Phe			(6) Glu
23	Leu	Val	Val			(7) Glu
24	Gln	Gln	Asn		(1) Gln	(8) Glu
25	Gly	Trp	Trp		(2) Gln	(9) Glu
26	Leu	Leu	Leu	Asp	(3) Asp	(10) Glu
27	$ValNH_2$	Met	Leu	Tyr (SO_3H)	(4) Tyr ((SO_3H)	(11) Ala
28		Asn	Ala	Met	(5) Thr	(12) Tyr (SO_3H)
29		Thr	Gln	Gly	(6) Gly	(13) Gly
30			Lys	Trp	(7) Trp	(14) Trp
31			Gly	Met	(8) Met	(15) Met
32			Lys	Asp	(9) Asp	(16) Asp
33			Lys	Phe-NH_2	(10) Phe-NH_2	(17) Phe-NH_2
34			Ser			
35			Asp			
36			Trp			
37			Lys			
38			His			
39			Asn			
40			Ile			
41			Thr			
42			Gln			

Fig. 10.2. Comparison of amino acid sequences of polypeptide hormones from the digestive system.

molecule is cleaved enzymatically several times to release big and little gastrins. Most of the biological activity of these gastrins resides in the four carboxy-terminal amino acids consisting of *Trp-Met-Asp-Phe-*NH_2. Several peptides that possess this terminal sequence have been shown to stimulate acid secretion. A synthetic pentapeptide (*pentagastrin*) that incorporates the terminal tetrapeptide sequence is frequently used for experimental studies.

The gastrin theory for control of acid secretion combines the observations that parasympathetic stimulation, acetylcholine (ACh), histamine and gastrin all cause acid secretion, whereas atropine (an anticholinergic drug), procaine (an anesthetic), sympathetic stimulation or certain antihistamines tend to reduce acid secretion under some experimental conditions. Presumably parasympathetic stimulation through release of ACh causes release of gastrin from the *G cell* in the antral mucosa. Gastrin travels via the blood to the fundus of the stomach where it stimulates the release of histamine from cells situated in the mucosa. Gastrin activates the synthesis of histidine decarboxylase, the enzyme responsible for histamine synthesis. Histamine in turn stimulates release of HCl from the parietal cell, probably via an adenyl cyclase-cyclic adenosine 3′,5′-monophosphate (cAMP) mechanism, resulting in decreased pH of the stomach contents. Gastrin may stimulate parietal cells directly without employing histamine as an intermediate, and some investigators conclude that histamine is not involved in endogenous acid secretion. Vagal stimulation or application of ACh may stimulate the parietal cells directly to secrete HCl.

During active digestion the pH of the stomach may be between 1 and 2. Low pH in the antral portion of the stomach (especially near the pyloric sphincter) reduces gastrin release.

Secretion of pepsinogen

The major gastric enzyme is the protease pepsin that is secreted by the chief cells in an inactive form, pepsinogen. Conversion of inactive pepsinogen to pepsin is accomplished by the presence of an excess of hydrogen ions supplied by HCl secreted from parietal cells. The optimum pH for vertebrate pepsins lies between 1 and 2, the normal pH range observed in the stomach following stimulation of acid secretion. The presence of acid on the surface of the gastic mucosa may activate a cholinergic reflex that evokes pepsinogen release. Parasympathetic stimulation via the vagus nerve causes release of pepsinogen, but hormonal control of pepsinogen secretion has not been established. A duodenal peptide motilin, has been implicated in regulating pepsinogen secretion. Gastrin causes release of pepsinogen only when applied in doses great enough to *inhibit* acid secretion by the parietal cells, implying that gastrin is not the normal factor causing pepsinogen release from the chief cells. Several other GI peptides can invoke pepsinogen release but do so only when applied in pharmacological doses.

Phases of gastric regulation

Gastric secretion is controlled via two levels. The *cephalic phase* involves stimulation of secretion via parasympathetic discharges elicited by the same stimuli that cause salivation; that is, sight, smell, taste, thought or presence of food. In the *gastric phase* of secretory control the presence of food in the stomach elicits secretion through vagovagal reflexes and/or through the gastrin mechanism. It has not been possible to determine which of these mechanisms is more important in controlling gastric secretion; probably all of these mechanisms operate in the normal digestive process.

Intestinal Events

While in the stomach, the bolus of food become saturated with the acidic gastric juices. These substances are thoroughly mixed by peristaltic contractions of the stomach to form an acidic, viscous fluid called *chyme*. Motility of the stomach, which is stimulated by the parasympathetic system, is responsible for churning the food mass and mixing it with gastric juices to form chyme. If the pH of the chyme in the antrum is sufficiently low (that is, less than 4.5) and of proper viscosity, the pyloric sphincter opens, and acidic chyme is squirted into the first segment of the small intestine, the *duodenum*. The exact mechanism controlling ejection of chyme from the stomach is not understood. Secretions also enter the duodenum from the exocrine pancreas and liver, through the *common bile duct*, in response to events taking place in the duodenum.

The duodenal mucosa contains many secretory cells, including cells responsible for digestive enzyme secretion as well as a variety of hormone-secreting cells and mucus-secreting cells. The intestinal mucosa is organized into thousands of tiny finger-like projections called *villi*. The presence of villi greatly increases the total surface area of the small intestine for both secretion and absorption of digestive products. The duodenum possesses more villi and secretory cells than do the more posterior sections of the small intestine (*jejunum* and *ileum*). The degree of specialization in the mucosa decreases progressively from anterior to posterior.

Secretion and pancreozymin

The presence of acidic chyme (pH less than 4.5) in the duodenum directly stimulates the *S-type cell* in the duodenal mucosa to release the peptide *secretin* into the blood. Bayliss and Starling discovered the existence of this factor that stimulates the pancreas to secrete basic juice and helps to neutralize the acidity of the chyme that has entered the small intestine. Although secretin levels in the blood do not increase following ingestion of a meal, the action of secretin on the exocrine pancreas is potentiated by another intestinal peptide hormone, *cholecystokinin*, that increases the blood following ingestion.

Originally it was believed that secretin was also responsible for stimulating secretion of the digestive enzymes normally present in the pancreatic juice, including the proteases chymotrypsin and trypsin, pancreatic lipase, pancreatic amylase and nucleases (DNase and RNase). After 40 years of controversy following the demonstration of secretin it was confirmed finally by Harper and Raper that purified secretin stimulates secretion of pancreatic fluid that is rich in sodium bicarbonate and poor in digestive enzymes. A second duodenal peptide, named *pancreozymin*, was found to contaminate some secretin preparations but not others, depending on the methods of preparation. Since *zymogen granules* represent vesicles of stored enzyme within the acinar (exocrine) pancreatic cell, the peptide that causes extrusion of zymogen granules from pancreatic acinar cells is logically called pancreozymin (pancreas-zymogen). It was postulated that the release of pancreozymin into the blood in response to the presence of peptides and amino acids in the chyme is due to direct actions of these molecules on pancreozymin-producing cells, similar to the action of hydrogen ions on the S-type cell. Sometimes the term *secretagogue* is applied to substances present in food, products secreted from the mucosa into the gut lumen or products of digestion that induce gastric or intestinal secretions.

Cholecystokinin

It was proposed in 1928 by Ivy and Goldberg that the presence of fat in the chyme or some of the digestive products stimulated release of yet another intestinal peptide that they named *cholecystokinin* (*chole* bile + *kystis* bladder + *kinein* move). Cholecystokinin travels via the blood to the gallbladder where it stimulates contraction of the smooth muscles comprising the walls of the gallbladder. At the same time it causes relaxation of the sphincter muscle that controls exit of bile from the gallbladder (the sphincter of Oddi). As a result, bile is expelled from the gallbladder, enters the bile duct (bile-bladder-move) and is transported to the duodenum. Bile is a viscous, complex mixture consisting largely of bile salts and bile pigments. Bile salts are powerful emulsifiers of fats (detergents). Bile pigments are breakdown products of hemoglobins, and they provide to bile and feces their characteristic colourations. Many other substances produced by the liver are present in bile, including metabolites of steroid hormones and inorganic iodide from deiodination of thyroid hormones. When bile enters the small intestine the bile salts emulsify globules of fat, causing them to be dispersed as small fat droplets within the aqueous digestive fluids. The emulsification of fats allows for marked reduction in the volume-to-surface ratio of the fat droplets, facilitating hydrolytic attack by pancreatic lipase to release glycerol and fatty acids for absorption.

Pancreozymin-cholecystokinin

Isolation and purification of GI peptides was finally accomplished by Mutt and Jorpes in Sweden more than half a century after the discovery of secretin by Bayliss and Starling. Investigators have since succeeded in identifying many distinct GI peptides. Secretin was confirmed as a separate peptide consisting of 27 amino acids. Purified secretin stimulated secretion of pancreatic juice that was low in enzyme content. However, the biological functions previously ascribed to PZ and CCK were found to reside in the same peptide consisting of 33 amino acids. Consequently the rather cumbersome name *pancreozymin-cholecystokinin* (PZCCK or CCKPZ) has been proposed to designate the single peptide that has been described in the literature under separate names for these separate functions. The *I-type* cell in the intestinal mucosa has been identified as the synthetic source for PZCCK.

The separate functional roles of secretin and PZCCK have been demonstrated elegantly in vitro with slices of exocrine pancreas. Physiological levels of PZCCK do not evoke release of basic pancreatic juice except in the presence of secretin,"' implying a permissive role. The action of secretin is markedly enhanced by PZCCK. Addition of purified PZCCK causes extrusion of zymogen granules from the acinar cells. Acetylcholine also causes extrusion of zymagen granules, suggesting a role for parasympathetic control over enzyme release from the exocrine pancreas. Parasympathetic influence and PZCCK probably operate via separate mechanisms. Atropine blocks the action of ACh but not of PZCCK. Purified secretin does not induce zymogen granule extrusion in these preparations. These data support the hypothesis for direct vagal (parasympathetic) influence over pancreatic enzyme release as being part of the normal regulatory mechanism; however, PZCCK does not require neural factors for its action.

Glucose-dependent insulinotropic peptide

Another intestinal GI peptide *enterogastrone*, was proposed in 1930 by Kosaka and Lim to be released in response to arrival of fat from the stomach. Enterogastrone was believed to inhibit gastric motility (peristalsis) as well as reduce acid secretion by the parietal cells. These actions would slow the entrance of fat into the small intestine and allow more time for proper processing of fat.

A peptide has been isolated from the intestine that inhibits gastric function. It was named *gastric-inhibitory peptide* (GIP) and is produced by the K cell in the duodenal mucosa. Chemically, GIP is similar to glucagon and secretin but is larger (42 amino acids). This peptide blocks gastric peristalsis and secretion of acid and enzyme. However, these effects occur only in the experimentally denervated stomach. It also stimulates release of insulin from the endocrine pancreas. This insulin-releasing action is now believed to be the physiological role for GIP. Following glucose uptake, GIP is secreted into the blood and causes release of insulin. Because of this action, GIP's name has been changed to *glucose-dependent insulinotropic peptide* while retaining the same acronym.

Neither bile secretion by the liver nor the secretion of *Brunner's glands* in the intestinal mucosa are influenced by GIP. Brunner's glands secrete viscid alkaline mucus that is believed to neutralize gastric juice entering through the pylorus. These glands are concentrated in the region of the pyloric sphincter and gradually decrease in frequency posteriorly, only occasionally being present in the most anterior portion of the jejunum. These cells also may secrete mucus in response to the presence of acidic, chyme in the duodenum.

Motilin

A unique small peptide (22 amino acids) that stimulates both motility and secretion of pepsinogen by the stomach has been isolated from the duodenum. This peptide, *motilin*, is structurally unlike the other GI peptides. Motilin is released from the *EC-type* cells when alkaline conditions are present in the duodenum, but the actual regulatory mechanism(s) governing its release is (are) uncertain. The physiological significance of this effect has been questioned since very alkaline conditions (pH 8.5-10) are required for releasing motilin. Acidification may cause motilin release under certain conditions in humans.

Phe--Val—Pro—Ile—Phe—Thr—Tyr—Gly—Glu—Leu—Gln—
Arg—Met—Gln—Glu—Lys—Glu—Arg—Asn—Lys—Gly—Gln

Fig. 10.3. Amino acid sequence for motilin.

Vasoactive intestinal polypeptide

Another peptide isolated from the intestinal mucosa has been shown to relax smooth muscle, thereby increasing the flow of blood to the viscera. This *vasoactive intestinal polypeptide* (VIP) consists of 28 amino acids and is structurally similar to both glucagon and secretin as well as to GIP. Vasoactive intestinal peptide is produced in neurons and by a pyramidal-shaped *H cell* found in the small intestine and in the colon of some species. Sufficiently high doses of VIP have been shown to produce secretin-like effects on the pancreas as well as a hyperglycemic response. These actions of VIP on the exocrine pancreas and blood sugar levels may be pharmacologic. A major role for VIP is that of an inhibitory neurotransmitter produced by intestinal nerves. It inhibits vascular smooth muscle but stimulates glandular epithelia. It is the inhibitory action on vascular smooth muscle that increases blood flow into intestinal tissues. The immunoactive VIP found in the brain is structurally identical to intestinal VIP.

Enterocrinin

In 1938 intestinal extracts were shown to stimulate intestinal secretion itself, and a hormone immediately was postulated in the usual manner to account for this activity. The purported hormone was named *enterocrinin*. However, both VIP and GIP have been shown to be capable of evoking secretion by the small intestine. Secretin and PZCCK are ineffective. Enterocrinin is probably not a descrete peptide but it remains to be demonstrated that intestinal secretion during normal digestion is controlled by either VIP or GIP or both.

Enteroglucagon

In addition to those hormones isolated from the small intestine, the *EG_I-type cells* produce *enteroglucagon*, which is structurally and functionally like pancreatic glucagon. Enteroglucagon is extractable in a small and a large form. The latter has been named *glicentin* and consists of 69 amino acids. This may be a prehormone form since it contains the glucagon sequence enclosed within peptide fragments of 8 and 30 amino acids. The physiological role for enteroglucagon and its relationship to pancreatic glucagon and glucose metabolism are open for speculation. A relationship to release of calcitonin (CT) and calcium regulation has been proposed.

Other gastrointestinal peptides

A peptide isolated from the duodenum of the pig specifically stimulates release of the inactive protease chymotrypsinogen. When activated in the duodenum, chymotrypsin enzymatically converts trypsinogen into the active protease trypsin. This peptide has been named *chymodenin*. This finding may represent a specialized refinement in control of specific enzymes from the pancreas according to the particular composition of the food mass.

A *gastrin-releasing peptide* (GRP) has been isolated from the antral gastric mucosa of dogs. It is composed of 27 amino acids and is structurally similar to *bombesin*, a peptide that may function as an enteric neurotransmitter.

Somatostatin is produced by neurons and paraneurons (D cells) located throughout the small intestine. It seems to function in paracrine fashion by inhibiting release of all other GI peptides. Somatostatin has been demonstrated in the arterial circulation of dogs following a meal and may act as a hormone to inhibit release of gastrin, insulin and pancreatic polypeptide. Two forms of somatostatin have been isolated. The first is identical to the hypothalamic tetradecapeptide neurohormone. The second consists of somatostatin plus an additional 14 amino acids. This may be a presomoatostatin form. *Dynorphin*, a heptadecapeptide from the central nervous system and pituitary gland is also present in the intestine. *Neurotensin* (tridecapeptide) and *substance P* (onadecapeptide) are neurotransmitters of the central nervous system that are also found in the intestine. Substance P was the first peptide to be identified in both the intestine and brain. It is not clear whether substance P and neurotensin are neurotransmitters only or if they have paracrine functions in the intestine.

The new biochemical methodologies for isolating and sequencing of peptides are allowing the isolation of new GI peptides faster than physiological studies can be performed to find a function for them! Porcine histidine isoleucine heptacosapeptide (PHI) is such a peptide. It consists of 27 amino acids, is structurally similar to the secretin family of peptides and is in need of a good, stable function.

Some proposed GI peptides

Numerous different peptides have been postulated to influence intestinal and gastric physiology, including *coherin* from neurohypophysial extracts, *villikinin*, *duocrinin*, *bulbogastrone*, *urogastrone*, *oxyntomodulin*, *sorbin* and several others from the gut. However, none of these proposed regulators has been shown to be an important, integral part of the digestive control mechanisms.

Phases of intestinal regulation

Three stages or phases of intestinal regulation can be identified involving neural and endocrine mechanisms. There appears to be a distinct *cephalic phase* mediated via vagal stimulation that influences pancreatic secretion. The *gastric phase* involves vagal and vagovagal stimulation of gastrin release that appears to influence pancreatic secretion. Finally, and certainly the most important regulatory mechanism, the *intestinal phase* relies primarily on release of peptides stimulated by the composition of the intestinal contents.

Chemistry of Mammalian Gastrointestinal Peptides

A number of specific peptides have been isolated from the GI tract of man and several domestic mammals. As described previously, these peptides exhibit the biological activities classically associated with secretin, gastrin and PZCCK and more recently with GIP, VIP, motilin and others. The sequences of amino acids that have been worked out for some of these peptides suggest that three chemical classes of GI peptides have evolved. The first group includes the gastrins, PZCCK and some related peptides. Secretin, enteroglucagon, VIP, GIP and PHI form the second group. The third group is composed of bombesin, GRP, substance P and a number of related peptides that have been isolated from amphibian skin (physalaemin, phyllomedusin, etc.) and molluscs (eledoisin). This last group are known also as tachykinins. Motilin and somatostatin are each unique peptides and are not similar to the peptides in any of these groups.

Embryonic Origin of Gastrointestinal Endocrine Cells

Although the use of immunological and fluorescent techniques has enabled investigators to identify the actual cellular sources for many of the GI peptides, there is still considerable disagreement with respect to the embryonic origin or origins of these cells in mammals. Pearse proposed that all of these GI cellular types as well as calcitonin-secreting C cells of the thyroid gland, parathyroid chief cells, α- and β-cells of the pancreatic islets, melanin-containing cells, adenohypophysial cells and the chromaffin cells of the adrenal medulla belong to the so-called APUD cellular series (amine content and amine precursor uptake and decarboxylation). These APUD cells are derivatives of neural crest or other neural ectoderm cells, suggesting that GI endocrine cells are of ectodermal rather than endodermal origin and that they have migrated into the intestinal mucosa early during development. An alternative view might be that at least some of these different cellular types have independently acquired APUD characteristics subsequent to or coincident with their differentiation from endodermal cells. The occurrence of certain GI peptides such as VIP in neural tissue argues strongly for a neural origin for these peptide hormone-secreting cells. A single origin for GI endocrine cells is supported further by observations that immunoreactive gastrin, CCK and glucagon appear to be localized in a single cellular type in the invertebrate chordate amphioxus and in the cyclostomes.

Complex Interactions of Gastrointestinal Peptides

Many studies have been published in the past few years that involve observation of the effects of administering combinations of GI peptides as well as the influences of one peptide on the release of another. Studies of this type indicate considerable overlap in the functional roles of the various peptides (for example, glucagon-like activity in secretin) although pharmacological doses usually were employed. At the present time it is difficult to sort out interactions due to structural similarities, pharmacological doses or both from those interactions that might represent true synergisms, functional overlaps or inhibitions. Consequently there is considerable literature concerning such interactions that will not be discussed here. The reader is encouraged to study literature and to find order in the chaos.

Influence of Gastrointestinal Peptides on Other Endocrine Systems

The influence of GIP on insulin release has already been mentioned, and it is possible that enteroglucagon release would also stimulate insulin secretion. Consequently GI peptides might influence the metabolism of carbohydrates, amino acids and fats after absorption.

Pentagastrin administered in small doses stimulates release of calcitonin from the C cells of the mammalian thyroid. Since experimental hypercalcemia produced by systemic infusion of calcium results in elevated levels of circulating gastrin, it has been proposed that uptake of calcium from the gut acting through the release of gastrin as a mediator effects release of CT. This increase in CT levels would enhance deposition of calcium in bones and offset any effects of parathyroid hormone (PTH) on this target tissue. The early release of gastrin during the processing of a meal may represent an "anticipation" of the consequent uptake of calcium that will occur so that CT released by gastrin lowers plasma Ca^{++}, thereby stimulating PTH release, which in turn would influence Ca^{++}/HPO_4^{-2} management by the kidney and indirectly enhance uptake of calcium from the gut. The action of PTH on bone would be blocked by CT. Enteroglucagon has also been suggested to evoke CT release in a similar manner.

Comparative Aspects of Gastrointestinal Peptides

There have been few studies concerning endocrine regulation of GI physiology in submammalian vertebrates, and most of these studies have been performed since the advent of purified mammalian peptides. Obviously the investigation into comparative regulation was hampered by the lack of understanding and interest in the general physiology of digestion in submammalian species. For example, although extensive research in teleosean fishes of commercial importance has been accomplished with respect to diets, growth and feeding ecology, few experiments have been concerned with physiological control mechanisms. The following brief account should serve to further emphasize the neophyte status of this area of comparative endocrinology and, it is hoped, encourage some developments in the study of comparative aspects of GI peptides.

Invertebrates

Immunoreactive gastrin has been extracted from the GI tract of two molluscan species. The levels of extractable gastrin are comparable to those of mammals. Gastrin has also been demonstrated in the neuroendocrine cells of an insect, *Manduca septa*, supporting possible neural origins for this peptide. These data suggest a broad phylogenetic distribution of these peptides associated with digestive functions. Peptides characteristic of invertebrate systems, are turning up in vertebrate guts and nervous systems. For example, the head inducing substance of *Hydra* has been found in human brain and intestine. The implications of these isolated observations may become more meaningful through studies of other invertebrate groups.

Class Agnatha: Cyclostomata

Cytological studies by Ostberg and co-workers on the intestine of the Atlantic hagfish *Myxine glutinosa* have revealed the presence of primitive open-type endocrine cells. These cells extend from the basal portion of the intestinal epithelium to border on the lumen of the gut. Hagfish intestinal endocrine cells do not possess APUD characteristics, although APUD-type cells have been reported in the pancreatic islets that differentiate into insulin-producing B-cells. Antibody to human gastrin, pentagastrin and porcine glucagon binds to endocrine cells in the *Myxine* gut. These intestinal endocrine cells in the Atlantic hagfish do not resemble zymogen cells either, suggesting a separate origin for the endocrine and enzyme-secreting cells. In contrast, the intestinal epithelium of larval and adult lampreys (*Lampetra* spp.) contains APUD-type cells that react to antibodies prepared against mammalian glucagon and gastrin.

Secretin-like and PZCCK-like activities have been demonstrated in intestinal extracts prepared from river lampreys, *Lampetra fluviatilis*, and sea lampreys, *Petromyzon marinus*. Both secretin and PZCCK activities were assayed by monitoring pancreatic secretions in the anesthetized cat. Similar observations have been reported for *M. glutinosa*. Gallbladder strips prepared from a Pacific hagfish, however, did not respond in vitro with contractions to porcine PZCCK although ACh caused contractions.. Secretion of intestinal lipase in this same species is stimulated by porcine PZCCK. These observations suggest that the evolution of gallbladder receptors for PZCCK occurred after the appearance of molecules in the intestine that possess PZCCK-like activities.

Somatostatin is not present in the intestines of hagfishes. It does occur in the pancreas.

Class Chondrichthyes

In their classical studies Bayliss and Starling reported the presence of secretin-like activity in extracts prepared from dogfish shark and skate intestines when these preparations were assayed in mammals. The activity they measured may have been due to secretin-like or PZCCK-like factors (or to both) present in these extracts. Intestinal extracts prepared from the holocephalan *Chimaera monstrosa* also possess PZCCK activity. Porcine PZCCK stimulates contractions in strips of gallbladder prepared from dogfish sharks, and the intensity of the response is proportional to the dose of PZCCK.

Class Osteichthyes: Teleostei

Only a few species of teleosts have been investigated within the entire class of bony Fishes. The first published observations are those of Bayliss and Starling who reported that intestinal extracts prepared from salmon (*Salmo salar*?) possessed secretin-like activity (possibly also PZCCK activity as well). Similar activities were reported for pike, *Esox lucius*, and cod, *Gadus morhua*, when intestinal extracts were assayed in either birds or mammals. The magnitude of these responses was similar to those induced by purified mammalian VIP. PZCCK activity has been reported in the intestine of the Atlantic eel, *Anguilla anguilla* and the pike. Isolated strips of gallbladder from Pacific salmon (*Oncorhynchus*) contract in the presence of porcine PZCCK, indicating sensitivity of the salmon gallbladder to the mammalian peptide.

A gastrin-histamine type of mechanism is present in the teleost stomach. Extracts from the gastric mucosa of sunfish (*Lepomis macrochirus*) stimulate acid secretion in bullfrogs, and large doses of histamine (10-15 mg/kg weight) induced acid secretion in the European catfish *Silurus glanis*. Histamine-induced acid secretion in cod (15 mg/kg) is blocked by certain antihistamines, supporting the existence of a mammalian-like regulatory system.

Somatostatin and motilin were not demonstrable in the intestine of *Gillichthyes mirabilis*.

Class Amphibia

Regulation of gastric mechanisms has been studied more extensively in frogs than in any other nonmammalian species. It appears that amphibians possess mechanisms very much like those of mammals, involving both neural and endocrine mechanisms. Stomachs of intact frogs or isolated gastric mucosa prepared from frogs (including *Rana pipiens*, *R. catesbeiana*, *R. temporaria* and *R. esculenta*) respond with acid secretion when subjected to ACh, histamine, pentagastrin or crude gastrin preparations from nonmammals or mammals. Similarly gastric mucosa isolated from a urodele, *Necturus*, secretes acid in response to pentagastrin. Treatment with atropine or surgical vagotomy reduces acid secretion in frogs as it does in mammals. Supposedly the release of pepsinogen in *R. esculenta* can be effected by increasing parasympathetic activity. Caerulein, a peptide previously thought to be present only in anuran skin and which is known to stimulate acid secretion from stomach mucosa in a variety of vertebrates, appears to be the endogenous "gastrin" in *R. temporaria*.

Bayliss and Starling reported that extracts from frog intestines would evoke pancreatic secretion in dogs, providing evidence for the presence of secretin-like or PZCCK-like factors or both. Frog gallbladders will contract in the presence of porcine PZCCK in vitro, which supports the possible existence of a PZCCK-like factor in amphibians as well as a role for PZCCK in regulation of gastric processes.

Class Reptilia

The only observation with respect to GI regulation in reptiles dates back to the observation by Bayliss and Starling that a factor or factors capable of causing pancreatic secretion in mammals is present in the intestine of a tortoise. Immunoreactive somatostatin is present in the intestine of the lizard *Anolis carolinensis*, but motilin is not. It is remarkable that additional studies have not bee reported or if reported have escaped recognition. Certainly the field is open for some careful comparative studies.

Class Aves

Mammalian gastrin can stimulate acid secretion in birds, and large amounts of PZCCK cause release of enzymes from the avian exocrine pancreas. However, no good evidence has been presented for a physiological role for these hormones in birds. Glucagon and GIP have no effects on pancreatic secretion. Extracts prepared from chicken intestines are strong stimulants of pancreatic secretion when assayed in turkeys, but these extracts are only weak stimulants in mammals (cat, rat). Purified porcine secretin only weakly stimulates exocrine pancreatic secretion in turkeys, but purified mammalian VIP is a potent stimulator. These data suggest that secretin-like activity in birds may reside in a molecule that is more like mammalian VIP than it is like secretin. However, chicken VIP has been isolated and differs structurally from porcine VIP at only four positions. Somatostatin and motilin have been demonstrated in the intestines of Japanese quail, but nothing is known about their functions.

11

Ovarian Hormones

A consideration of physiology of the reproductive tract should be approached with a somewhat different attitude than is generally accorded most of the other organ systems of the body. The hormones secreted by the reproductive tract are not vitally essential for the well-being of the individual. In fact, any part or all of the structures directly concerned with reproduction may fail to function or be surgically removed without affecting general health or life expectancy. Therefore, reproduction is essential only for the propagation of the species, whereas to the individual it is merely a privilege which may or may not be indulged.

The relative independence of the reproductive process has made it possible for widely different adaptive mechanisms to occur without marked interference with the general economy of the body. This seems so particularly in the light of the diverse developments that have occurred in the course of the evolution of viviparity. However, it seems logical to assume that the earliest adaptations to appear were those associated with follicular development and ovulation. What these were is difficult to determine, but apparently the simplest arrangement found at the present time among the vertebrates is one involving the interaction of the pituitary gonadotropins and estrogen of the Graafian follicle. This appears to be the basic situation in oviparous species.

Therefore, if this be so, it seems reasonable to conclude that the reproductive cycles of all vertebrates are physiologically the same through follicular development and ovulation. That is, the pituitary-ovarian interactions responsible for follicular growth and ovulation in oviparous species are homologous with those of the follicular phase of the estrous cycle in mammals. Although this probably is the most primitive situation found among the vertebrates, it represents in itself a specialization, the antecedent of which is not clear. Evidence drawn from the protochordates is limited mostly to discussions of possible morphological homologies, and supporting physiological information is lacking or unconvincing. A more extensive physiological and biochemical consideration of the lower chordates would probably contribute much toward the solution of this problem, but at present fish constitute the oldest group for which we have reliable experimental data.

A brief consideration of the endocrine adaptations in fish for pituitary-ovarian function and viviparity can serve as a useful introduction to a discussion of the more specialized situations found in mammals. It seems that both FSH and LH are secreted by the fish pituitary and probably LtH, the luteotropic hormone or prolactin. The ovaries secrete estrogen, but there appear to be wide variations in amount. The estrogenic activity of extracts prepared from teleost ovaries is low, whereas that of similar preparations from elasmobranch eggs and ovaries is surprisingly high. It is also of considerable interest that estradiol-17β has been obtained from the ova of *Squalus suckleyi*, and both estradiol and a small amount of estrone are present in extracts of whole ovaries.

That the theca interna secretes estrogen in the mammalian ovary is an opinion that seems widely accepted, though it is based in large measure on circumstantial evidence. Almost nothing is known about this for the lower vertebrates. In many kinds of animals there is an association of nurse cells with developing ova, and the granulosa of the vertebrate follicle seems to perform a similar function. Also, the variety of substances deposited in the yolk makes it quite unlikely that the granulosa is metabolically unacquainted with steriod compounds regardless of their origin. In view of this and the fact that the granulosa is responsive to estrogens the theca interna may represent an adaptive mechanism, the original function of which was stimulation of metabolic activity of the granulosa and deposition of yolk in the ovum.

If this thought seems of value it is possible to continue with certain implications. The first of these is the possibility that a functional theca interna arose in conjunction with the origin of pituitary-ovarian interactions. This would imply that the importance of estrogen in the physiology of the ovary appeared quite early in vertebrate evolution and was preceded by a condition in which ovarian function was regulated by other means than those involving the action of estrogen. It also seems probable that the ovarian adaptations appeared before other organs of the body gained competence to respond to estrogenic action. This applies particularly to specialization of the mullerian ducts for the transport of large ova, secretion of egg coats, and retention of young under conditions of viviparity.

Evidence in support of these ideas is not abundant, but what there is appears consistent and suggestive. The small amount of estrogen in teleost ovaries in contrast with the high concentration in the ova and ovaries of oviparous elasmobranchs is important, and even more so is the fact that the hormone in the ova of *Squalus suckleyi* is estradiol-17β which also is the follicular hormone of mammals. However, the amount of estrogen actually stored in the egg may itself be of an adaptive nature and as a result show wide variation among species that have large yolk-laden ova. A possibility in the dogfish is that estrogen in the developing ova may help maintain the large hyperemic uteri during the period of gestation between ovulation and development of the next crop of follicles. An observation that supports this is that the yolk sacs of uterine young after more than twelve months gestation yet contain a surprising amount of estrogen. Even so, our general lack of information regarding the evolutionary aspects of follicular development is emphasized by the fact that with the single exception of *Squalus suckleyi* the estrogenic compounds secreted by the ovaries of vertebrates, other than mammals, have not been identified, nor has a comparative study of follicular structure been made.

Corpora Lutea

The chief physiological difference between reproductive cycles of ovipara and the estrous cycles of mammals is the presence of corpora lutea that are capable of secreting progesterone. Thus, the estrous cycle is composed of two parts, first, a follicular phase characterized by the development of one or more Graafian follicles and which terminates at ovulation, and, second, a luteal phase during which the ruptured follicles are converted into corpora lutea. However, the presence of corpora lutea in the ovary is not a distinctive mammalian feature. Luteal bodies have been found in representatives of every class of vertebrates and certain protochordates, and are present in both ovipara and vivipara.

Corpora lutea may be formed either as a result of follicular atresia or organization of the follicle subsequent to ovulation, and their probable function and evolutionary significance have received considerable attention in several recent general discussions. It appears quite obvious that corpora lutea in the lower vertebrates not only are not restricted to vivipara but that successful gestations of long duration have been established in the absence of functional corpora lutea. It has been found that hypophysectomy of *Mustelus canis*, a viviparous dogfish, soon after the ova have entered the uterus, does not prevent normal development of the embryos during the following three and one half months. Also, the involution of follicles following hypophysectomy showed that corpora lutea atretica can be formed in the absence of the pituitary.

These observations on *Mustelus canis* are very much like those by Bragdon (1951, 1952) on two species of ovoviviparous snakes, *Thamnophis sirtalis* and *Natrix sipedon*. He emphasized the similarity in appearance of corpora atretica and post-ovulation corpora lutea and showed by hypophysectomy and castration that they were not essential for normal gestation. He also found that hypophysectomy brought about atresia of the large follicles and formation of corpora lutea. Thus, it is quite evident that the corpora lutea in these two species of snakes and in *Mustelus canis* are not concerned with gestation nor are they a result of pituitary luteinizing action.

One thing held in common by all vertebrates that have yolk-laden ova is the development of luteal bodies, both as a result of atresia and subsequent to ovulation. This is accompanied by the transformation of the granulosa into phagocytic foam cells that actively ingest yolk of moribund ova and detritus remaining in ruptured follicles. Phagocytic activity of the granulosa has been reported in a wide variety of vertebrates including mammals. Therefore, it seems not an unlikely possibility that with the development of polylecithal ova this reaction of the granulosa arose as an adaptive device for disposing of large moribund ova and tidying up the sizable follicular cavity after ovulation. This is in agreement with the instances cited in which corpora lutea do not function as endocrine glands of gestation, a condition probably true of all viviparous anamnia and all but a few doubtful exceptions among live-bearing reptiles.

If this be so, then it is clear that the next stage in the evolution of viviparity leading to conditions found in mammals was the adoption of the corpus luteum into the family of endocrine glands. When and how this took place is, of course, a matter of conjecture, but a good guess might be that it first occurred in the reptilian stock that gave rise to mammals. That a similar condition exists in present-day viviparous reptiles is a possibility, but the evidence brought forth in support of this has been, so far, rather unconvincing. Our knowledge, however, of the physiology of the mammalian corpus luteum gives some basis for assumptions regarding steps that might have been involved in the process of acquiring endocrine status.

In higher mammals a corpus luteum is formed by the action of the pituitary luteinizing hormones (LH), and it secretes progesterone in response to pituitary luteotropic hormone (LtH or prolactin). Therefore, the height of specialization of the mammalian corpus luteum has involved acquiring the capacity to respond to two pituitary hormones as well as the secretion of progesterone. This does not necessarily imply that all three of these properties were gained simultaneously, and it seems reasonable to suspect this was not so. It would appear more likely that during the deposition of yolk and subsequent phagocytosis associated with follicular atresia and ovulation, the cells of the granulosa might have become quite accustomed to handling a variety of steroids, either as secretions or degradation products. As to how progesterone became the dominant steroid produced, we have no positive evidence, but it may be of some significance to mention that cholesterol is present in the yolk, and when it is used as the precursor for synthesis of steroid hormones, progesterone is one of the first derivatives.

Therefore, it is suggested that synthesis of progesterone, probably on a small scale, was the first step in the specialization of the corpus luteum, and, second, this process was augmented by the acquisition of competence to respond to the pituitary luteotropic hormone (LtH). These two processes most likely became well established in primitive mammals whose ova yet contained considerable yolk, a condition similar to, though less advanced than, present-day monotremes.

The development of corpora lutea in monotremes indicates a definite response to luteinizing hormone (LH), so this adaptation also occurred before the loss of yolk from the ovum. This is of particular interest, since the cells of the granulosa of such follicles are yet capable of phagocytosis. Although large atretic follicles may burst and their contents absorbed in the interstitial spaces of the ovary, the granulosa of those that do not burst ingest the degenerate contents of the ovum. It appears that both here and in the lower vertebrates, yolk or debris in ruptured follicles is required to elicit the phagocytic reaction of the granulosa, as small follicles in which vitellization is not advanced are reabsorbed directly. In the lower vertebrates large corpora lutea are the result of atresia of large follicles, or organization of follicles from which large ova have been discharged at ovulation. The significance of this may be that with the decrease and ultimate loss of yolk in the higher mammals the action of LH became more important in the development of corpora lutea.

Thus, in some such way as this, the corpus luteum was brought into the endocrine system and the luteal phase of the estrous cycle was established. This evidently occurred in ovipara and was not necessarily associated with viviparity. It may not be entirely fanciful to think of the primitive endocrine function of the corpus luteum as being one of facilitating the old reptilian custom of carrying ova in the uterus until a favourable time and place was found to lay the whole clutch at once. It is an exciting thought that this may have developed into a timing device for laying, or in a sense parturating, ova in coordination with the involution of the corpus luteum. In fact, in *Ornithorhynchus* regression of the corpus luteum begins before the egg is ready for laying. A matter of equal interest is that during retention of the eggs in the uterus of both the duckbill platypus and spiny anteater (*Echidna*) embryonic development proceeds to a stage approximately equivalent to that of a chick embryo at 40 hours incubation. The exact length of time the eggs remain in the uterus is not known in either instance, but it is probably about two weeks.

This modest gesture toward viviparity in monotremes is in some respects reflected in marsupials. Gestation is approximately of the same length as the luteal phase of the estrous cycle. Luteal action following ovulation in the unmated animal gives rise to a condition known as pseudopregnancy in reference to its resemblance to true pregnancy. Actually the physiology of these three conditions is the same, since each depends upon luteal secretion in response to the pituitary luteotropic hormone and each is terminated by involution of the corpus luteum.

The salient features of reproduction in marsupials are well represented by those seen in the Virginia opossum (*Didelphis virginiana*), which has been studied more thoroughly than others. It is polyestrous during the breeding season, the cycles being about 28 days in average length, and gestation is about $12^1/_2$ days. Theoretically, this animal could produce a litter of young at the conclusion of each estrous cycle without disturbing the estrous rhythm. This, however, is prevented by the stimulus to the teats when suckling young are in the marsupium, which brings about postponement of estrous cycles for the duration of lactation, a period of something over three months. The dependence of lactation in higher mammals upon the pituitary luteotropic hormone and the modifications that occur in pituitary- ovarian function during lactation make these observations in marsupials of particular interest. For instance, are these adaptive interactions also present in lactating monotremes? If so, what was the evolutionary sequence of their appearance? It is quite evident that the pituitary hormone referred to as luteotropic hormone or prolactin is very old and was performing hormonal functions long before the advent of progesterone-secreting corpora lutea, and certainly before mammary glands came into being. If it is true that luteal function first appeared in viviparous reptiles, then, of course, it becomes obvious that corpora lutea came under pituitary control before mammary glands.

The evolutionary trend of viviparity beyond the condition found in marsupials was that of developing means for lengthening the period of gestation. Our knowledge of the methods that were devised is limited to a comparatively few species, but, judging from these, the first adaptations were concerned with prolonging the functional life of the corpus luteum. This seems to hold for all mammals whose gestation period is longer than the luteal phase of their estrous cycle and is associated with specialization of the chorion in the development of various methods of placentation. This took the form of adoption of endocrine functions by the chorion, and probably the first of these to be acquired was the ability to secrete luteotropin.

The chorionic luteotropins that are known are proteins or peptides, and although they are capable of prolonging the functional life of corpora lutea in their respective mammalian groups, they differ from each other and from the pituitary luteotropin in both their chemical characteristics and physiological properties. These differences suggest that chorionic luteotropins may be parallel adaptations that appeared after the mammals in which they are found had become differentiated from the common parent stock.

It is quite possible that other means than the action of chorionic gonadotropins might have been developed for lengthening the functional span of the corpus luteum. Information regarding the endocrinology of gestation, even for most laboratory and domestic animals, is little more than elementary and, at best, scarcely suitable for the present discussion. However, there are a few notable exceptions, and, although they do not represent a direct progression in the sense that one situation gave rise to another, they do show how the problem of lengthening gestation has been solved independently in a few distantly related groups of mammals.

The placental adaptation described for the rat probably represents one of the simplest methods for prolonging luteal function. Stimulation of the uterine cervix of a rat in estrus induces pseudopregnancy which lasts about 12-14 days. This condition, normally a result of mating, is identical with that of the first two weeks of pregnancy and is supported by the action of the pituitary luteotropic hormone on the corpora lutea, causing the secretion of progesterone. The length of normal pregnancy, however, is 21-22 days and the problem involved is the continuation of luteal function for about a week beyond the period of pseudopregnancy.

Corpora lutea in rats are essential for a successful gestation and if the ovaries are removed, especially during the first half of pregnancy, resorption or abortion occurs. Also, hypophysectomy before the eleventh or twelfth day causes death and resorption of the embryos, but, if performed later, gestation continues to term. It seems significant that the eleventh day is also when mesoderm and allantoic blood vessels grow into the ectodermal trophoblast, as it suggests that the time at which the pituitary can be removed without terminating pregnancy is correlated with placental development. These observations show that the pituitary luteotropin is not required for gestation after the eleventh day and indicate that a luteotropin from some other source may be present.

Astwood and Greep (1938) found that extracts of rat placentae contained a luteotropic substance. This was demonstrated by the ability of such preparations to continue luteal function for the production of deciduomata in hypophysectomized rats. Averill et al. (1950) were able to maintain pregnancy in rats hypophysectomized on the sixth day by implanting placental tissue, and Ray et al. (1955) have shown that the rat's placenta contains hormonal activity comparable with that of pituitary mammotropin (prolactin). These observations show clearly that the rat's placenta contains a hormone capable of sustaining luteal activity.

There is certain evidence, however, that seems to indicate that the placental luteotropin in rats may act only as a kind of double surety for the success of gestation, since, under certain circumstances, the last half of pregnancy may continue in the absence of corpora lutea. Haterius (1936) found that after removal of all the foetuses save one, the placentae being left in place, castration did not prevent the remaining foetus from going to full term. Zeiner (1943) reported that abortion may not occur if ovariectomy is done with care in two stages, one ovary being removed on the thirteenth and the other on the fifteenth day. Then, it also has been shown that the functional life of corpora lutea in the nonpregnant animal may be extended to 19 or 21 days when both uterine horns contain massive deciduomata.

The endocrine adaptations for gestation in mice are similar to those in rats. Castration at any time during pregnancy precipitates resorption or abortion. Hypophysectomy at mid-term does not interrupt gestation. The observations of Gardner and Allen (1942) strongly suggest that the placenta might secrete a luteotropic hormone. Removal of the pituitary ten days after copulation produced involution of the adrenal cortex and follicles and interstitial tissue of the ovary, yet luteal function apparently continued. It is significant in this respect that the mammary glands continued to develop and contained milk at parturition, and the pubic ligaments lengthened as in normal pregnancy. These effects not only indicate the presence of a luteotropin, progesterone, and relaxin, hormones adapted for gestation, but the development seen in the mammary glands also indicates an additional similarity between placental and pituitary luteotropin (prolactin).

Forbes and Hooker (1957) tend to disagree with tins view. They conclude from analyses of the progestin content of the blood of pregnant mice, and the cytology of the corpora lutea, that progesterone is probably secreted by the placenta during the last half of gestation. They did not, however, test the effects of castration on blood progestin, nor does this thought seem in keeping with the fact that hypophysectomy on the tenth day does not interrupt pregnancy while castration at this time usually does. Also, the evidence they cite for placental secretion of progesterone is drawn from other species, primates and the mare, in which endocrine adaptations are quite different from those in the mouse. Since progestational activity, as determined by the Hooker-Forbes method, has been found in blood and tissue extracts of a wide range of vertebrates, both ovipara and vivipara, and is not necessarily associated with luteal function or gestation, some doubt may be expressed as to the value of such data for a discussion of the evolution of viviparity other than demonstrating that progestins were components of the steroid milieu long before progesterone became of importance as a hormone of pregnancy.

The secretion of a luteotropin by the placenta probably arose independently as an adaptation in many distantly related groups of mammals. The diversity of hormonal adaptations suggests this possibility. The point at issue, of course, in all instances is the continuation of luteal function, and consequently gestation, beyond the normal length of the luteal phase of the estrous cycle. In certain species, however, pregnancy lasts for many months, and this has been made possible by an additional series of adaptive mechanisms for a further continuation and maintenance of a favourable hormonal situation. These arrangements evidently have taken widely different forms and it may be helpful to discuss first those about which most is known before attempting to make comparisons.

One of the trends in the evolution of viviparity beyond conditions described for the rat and mouse, has been in the direction of further specialization of the placenta as an endocrine organ. This has progressed to a point in primates at which the placenta not only secretes a luteotropic hormone, but also elaborates estrogens and progesterone. In fact, it has attained physiological autonomy to the extent that it can maintain itself in the uterus in the absence of maternal pituitary and ovarian hormones.

The physiology and timing of the successive events that occur in primates during gestation has been studied most extensively in the rhesus monkey, *Macaca mulatta*. From the standpoint of comparative endocrinology these events can be considered as belonging to three different stages or periods of gestation. The stages overlap to a certain extent and, thus, form a sort of endocrine relay which passes along the care of the developing ovum from the time of ovulation to parturition. The first of these is identical with the luteal phase of the menstrual cycle, the second is the period during which secretory activity of the corpus luteum is prolonged under the influence of the chorionic luteotropin, and the third begins with the attainment of physiological autonomy by the placenta.

Implantation of the free blastocyst begins in the rhesus monkey on the ninth and tenth days following ovulation and a solid trophoblastic plate which invades the uterine epithelium is formed. Between the eleventh and fifteenth day the trophoderm differentiates into a reticular mesh containing syncytial trophoblast, the lacunae of which are filled with maternal blood, and between the fifteenth and thirty-fifth day chorionic villi develop and the definitive placenta is formed.

These stages in placentation have several important correlations with the three endocrine stages of gestation mentioned above. The first of these is the fact that development of trophoderm is well advanced before the end of the cyclic luteal phase which ordinarily would terminate on approximately the fourteenth day. It is of interest that chorionic luteotropin is present in the blood as early as the

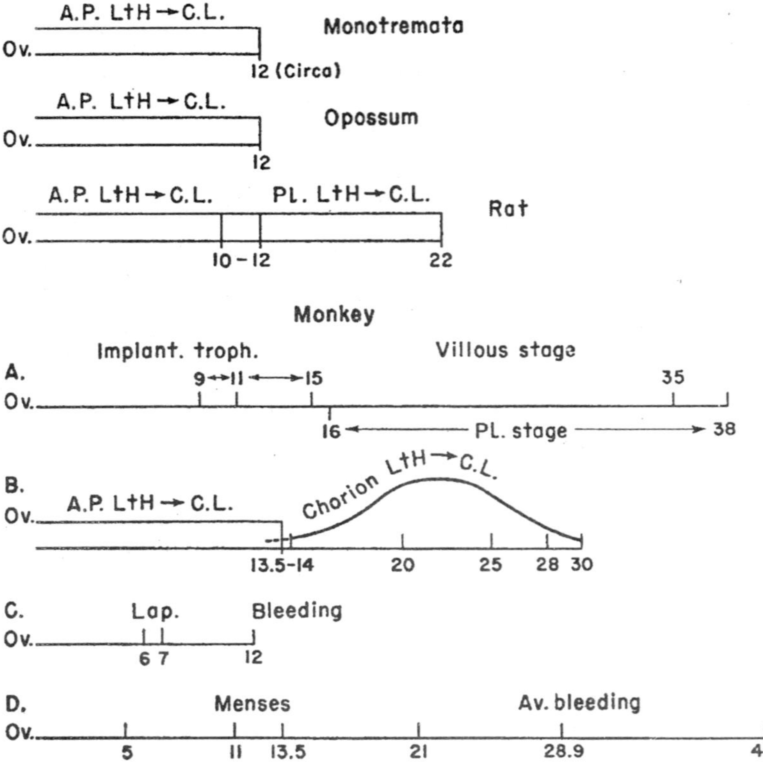

Fig. 11.1. A schematic comparison of hormonal activities preceding egg-laying in monotremes, gestation in the opassum and rat, and early pregnancy in the monkey. A—Development of the placenta. B—Endocrines of early pregnancy. C—Pituitary luteotropin (prolactin). D—Effects of chorionic gonadotropin on the length of normal menstrual cycles.

fourteenth day after ovulation, reaches a maximal concentration by the end of the following week, and is absent after about 30 days. Thus, it is seen that the presence of the chorionic luteotropin is correlated with placental development.

These observations raise a number of questions regarding endocrine function. One of these is the significance of the replacement of the pituitary luteotropic hormone (prolactin), which has maintained luteal secretion for two weeks following ovulation, by a chorionic luteotropin. Or, one may ask why it is that a corpus luteum involutes at the end of a corresponding length of time in a normal menstrual cycle? Hisaw (1944) found that injections of 120 to 240 I.U. prolactin daily did not postpone menstruation, while the average functional life-span of the corpus luteum was about doubled when 500 to 100 I.U. of human chorionic gonadotropin (Prolan B, HCG, PU) was given daily. This seems to indicate that the corpus luteum in monkeys has a longer life expectancy when maintained by placental than by pituitary luteotropin.

A point of equal interest is that the placental luteotropin in primates stimulates the corpus luteum to secrete both estrogen and progesterone, a capacity not possessed by the pituitary luteotropin. Also, under these conditions, a corpus luteum of the menstrual cycle develops the characteristic morphology of a corpus luteum of pregnancy. However, in nonpregnant monkeys, treated with human chorionic luteotropin, involution of the corpus luteum eventually sets in and the animal menstruates about 30 days after ovulation. The indications are that menstruation is due to failure of the corpus luteum to secrete progesterone, while the secretion of estrogen may continue, and, when this capacity is lost, a second bleeding is brought on by the withdrawal of estrogen. From this it seems that the function of the chorionic luteotropic hormone in the rhesus monkey is that of prolonging the functional life of the corpus luteum until the placenta is established, and has acquired the capacity to secrete progesterone and estrogens.

This view of placental physiology is supported by substantial experimental evidence. Removal of the corpus luteum as early as the twenty-fifth day does not interrupt gestation. The placenta is retained in the uterus in the absence of both ovaries and fetus and is parturated at term. Hypophysectomy as early as the thirty-second day does not interrupt pregnancy.

The hormonal adaptations associated with pregnancy in the rhesus monkey are probably common primate features. It is well established that in women removal of the corpus luteum or the ovaries alter placentation does not cause abortion. Recently Little and co-workers (1958) reported the continuation of gestation and birth of a viable infant following hypophysectomy. The pituitary was removed during the twenty-sixth week of pregnancy and delivery was induced at the end of the thirty-fifth week. During this time there was also a continuation of the excretion in the urine of approximately normal amounts of estradiol, estrone, estriol, pregnanediol, and chorionic luteotropin.

Chorionic luteotropin was discovered first in the urine of pregnant women and later in the rhesus monkey and chimpanzee. It appears in each during placental development, but there is a marked difference in the amount excreted and the length of time it is present. The amount excreted by the monkey and chimpanzee is very small in comparison with that of women.

The chorionic luteotropins of these three primates are apparently the same, as they possess the same chemical and physiological properties. The hormone will not promote follicular growth in either monkeys or women or in hypophysectomized rats but will luteinize existing follicles. It also stimulates secretion of both estrogen and progesterone by the corpus luteum and will prolong luteal function. This probably represents independent responses of the two components of the corpus luteum derived from the theca interna and the granulosa. Corpora lutea of rats lose their capacity to secrete progesterone within a few hours after hyphophysectomy yet they may remain in the ovaries for months and will secrete estrogen in response to human chorionic luteotropin. The reports by Westman and Jacobsohn (1937), Robson (1937) and Selye et al. (1935) of the beneficial effects estrogen has on corpora lutea encourage the thought that it may contribute to the prolongation of luteal function.

The mare is the only other mammal for which there is sufficient information regarding the hormones controlling reproduction to be useful in a discussion of comparative endocrinology. The endocrine mechanism for extending luteal function in the mare beyond the limits of the estrous cycle is in some respects like that described for primates. A placental hormone appears in the blood with nidation of the blastocyst on about the fortieth day, reaches a maximum concentration between the fiftieth and one hundredth day, and is absent after about the one hundred and seventy-fifth day. This hormone differs from the placental luteotropin of the rat and of primates in that it is a strong promoter of follicular development as well as having luteinizing capacity. Its presence is associated with the development of numerous accessory corpora lutea which disappear by the one hundred and eightieth day, and from this time until parturition the ovaries contain neither corpora lutea nor large follicles.

There is some question regarding the exact source of the placental hormone of the mare, since it appears to be formed in the tissues of the endometrial cups. Also, it is not clear whether the hormone induces secretion by the corpora lutea, and in this sense is a luteotropin, or whether its function is more that of a gonadotropin which furnishes accessory corpora lutea on which the animal's own hypophyseal luteotropin can act. It seems possible, of course, that it could take part in both of these activities. But, regardless of the hormonal interactions at this stage of gestation the outcome is a continuation of luteal function until the placenta is fully established and apparently becomes an autonomous structure. The independence of the placenta during advanced pregnancy is shown by the fact that ovariectomy does not interrupt gestation or the excretion of estrogen.

It is of importance to note that if pregnancy is longer than the luteal phase of the estrous cycle, provision is made for the secretion of estrogen as well as progesterone. The estrogen present during pregnancy in the mouse, rat, and rabbit apparently is derived from the ovaries, whereas in the primates and the mare, the placenta is the source. It has been shown that small doses of estrogen greatly facilitate progesterone in the maintenance of pregnancy. Also, the amount of estrogen present in mice, rats, and rabbits during pregnancy is small in comparison with that secreted by the placenta of primates and the mare. This difference is correlated with the observation that small amounts of estrogen will inhibit progesterone in the promotion of a decidual reaction in rats or a progestational reaction in the endometrium of rabbits. In the monkey, and presumably other primates, large doses of estrogen not only do not inhibit an effective dosage of progesterone on the uterus, but actually also take part in a synergistic reaction that greatly intensifies the progestational effects.

Although the progestational hormone produced by the placenta in certain mammals, so far as known, is progesterone, variations occur with respect to the production of estrogens. Estradiol-17β and estrone are probably the principal estrogens of the human menstrual cycle, although some estriol is present. As pregnancy advances, there is a progressive increase in the amount of estrogen excreted in the urine and there is also a large amount of estrogen in the placenta. This also is probably true of chimpanzees, since estradiol-17β, estriol, and progesterone have been obtained from their placentas. The secretion of estrogens by the primate placenta is well established, but a point of particular interest is the relatively large amount of estriol produced, which raises a question as to whether it is derived entirely as a metabolite of estrone. Although estriol is excreted following the injection of estrone in human beings, the amount excreted is not influenced by the presence or absence of a corpus luteum, nor by pregnancy. Also, such conversion does not occur when estrone is incubated with human tissues, nor is estriol obtained when the placenta is perfused with estradiol and estrone. However, regardless of these questions, the relatively large amount of estriol present during gestation, and the fact that it is peculiarly a primate steroid, suggest that it, as well as progesterone, probably should be considered a hormone of pregnancy.

Evidence for the placental origin of estrogens in the mare is equally if not more convincing. The dominant placental estrogens are equilin and equilenin, and they are present only during pregnancy. Another peculiarity is that they attain their highest concentration in the blood after the serum gonadotropin has disappeared. These and similar observations that have been discussed for the rhesus monkey indicate that secretion of placental luteotropins and steroid hormones is not evoked by influences from outside the placenta but that these adaptations are designed for completely independent action.

Somewhat in this connection is also the question as to why there should be several different estrogens. It may be significant that they do differ widely in their capacity both as to induction of fluid imbibition by the uterus and stimulation of uterine growth, some being much more effective for one than the other. It also is of interest that in these processes under normal conditions a strong estrogen is invariably associated with a weaker one and that under certain experimental situations it has been shown that one may influence the action of the other. The physiological meaning of this is not clear but it seems possible that some estrogens might in some way be better suited than others for maintaining uterine conditions during gestation.

The general trend in the evolution of viviparity in mammals seems to have been as described in the present discussion, but this opinion is based on a comparatively few species that have been studied more thoroughly than others. Emphasis has been placed on these with the thought that their endocrine adaptations may show a basic plan that might be found useful in making comparisons with similar specializations in species for which the endocrine situation is less well understood. This seems of particular importance for consideration of the diverse endocrine adaptations that have occurred in the specialization of the placenta as an endocrine organ. For instance, in rabbits the corpora lutea are essential for gestation and are the only source of progesterone, but whether luteal function beyond the normal limit of pseudopregnancy is maintained by a placental luteotropin, as in rats and mice, has not been conclusively demonstrated. Among ruminants, removal of the corpus luteum produces abortion in cattle, but in the ewe ovariectomy on or after the fifty-fifth day of pregnancy does not always result in abortion, and the concentration of progesterone in the blood continues as in pregnant sheep with intact ovaries. It is suggested that in sheep the placenta is the major source of progesterone during the last two trimesters of pregnancy. This apparent independence of the placenta resembles the condition in the mare, but the presence of a placental luteotropin at any time during pregnancy is unknown.

Many similar instances in which the placenta seems to function as an endocrine organ could be cited, and the impression gained is that each represents a series of adaptations of independent origin that are peculiarly suited for the functional needs of a particular group of mammals. There is no obvious evolutionary sequence, but rather, divergence into what seems to have been parallel lines of development. Seen at the present time, specific situations appear unrelated, except in the general sense that each represents attainment of partial or complete physiological autonomy of the placenta. Indeed, the hormones elaborated by the placenta may differ both physiologically and chemically and be restricted to the group of mammals in which they are found. This is particularly true of the placental luteotropins as seen in the rat, mare, and primates and also seems to apply to such estrogens as estriol, equilin and equilenin. Our limited knowledge is such that it is highly probable that many additional placental adaptations remain to be discovered, but from what is known it also appears that in the evolution of viviparity the acquiring of endocrine function by the placenta furnishes notable exceptions to the generalization which states in effect that *it is not hormones which have evolved but the uses to which they are put*.

12

Hormones in Mammalian Population

This paper reviews evidence pertinent to the theory that physiological, especially endocrine, responses to population density constitute a feedback mechanism which regulates the growth of mammalian populations. It is generally assumed that population growth is regulated directly by the availability of essential needs, such as food. If this assumption were correct a population would not stop growing until its food was exhausted, with a threat to the survival of the species. Critical studies show that, other than locally, acute shortages of environmental needs seldom occur. Many populations cease growing or decline with adequate supplies of food, harborage, and other necessities. Usually the subordinate animals are forced competitively into disadvantageous situations with regard to obtaining food or to predation. Similar statements apply to the role of disease in population declines. Evidence of epidemics decimating populations, except locally, is rare. Disease may be a factor increasing mortality during a population decline, but probably it is usually secondary to other factors, such as diminished resistance. Finally, striking declines in populations occur that have no obvious explanation.

Populations of rats can be reduced by decreasing the capacity of the habitat in terms of food and cover, but starvation does not occur. These declines are effected by competitive situations which decrease infant survival and increase mortality of the subordinate rats. Furthermore, most declines in numbers, once begun, continue to well below the point to which any lowering of the environmental capacity or disease epidemic should have reduced them, and the rate of decrease is greater than would be expected. These facts point to the existence of some underlying mechanism which regulates the size and growth of mammalian populations.

It seemed that any inclusive explanation for the regulation of growth of natural populations would include an intrinsic adaptive system through which environmental factors would act. Social pressure in the form of behavioural interactions is always present and was believed to be the key factor in regulating population growth, even though species vary considerably in their social organizations. It was thought that social competition could regulate population development by eliciting physiological adaptive responses in some proportion to population density. Increased competition therefore should stimulate pituitary-adrenocortical activity and decrease reproduction and resistance in some proportion to density. These responses to social competition would be density-dependent and not geographically, ecologically, or specifically limited. This feedback theory implies that environmental exigencies would act by increasing social competition. The validity of this theory has been tested for a few species of rodents in the laboratory, and evidence consistent with it is accumulating from the field, although it is recognized that specific differences will probably appear.

The balance of this paper summarizes evidence from laboratory and field that (1) social competition produces adrenocortical, reproductive, and other endocrine responses in relation to population density and (2) these responses regulate population growth through their effects on reproduction and mortality.

Experiments with Populations of Limited Size

A basic question is whether or not interactions between animals can provoke responses of the adrenal cortex and reproductive organs in some proportion to density. Measurements of reproductive function and adrenal weight have been used as indices of the degree of adaptive response in a population. It is assumed here that adrenal hypertrophy indicates increased adrenocortical function.

Vicious fighting results from introducing a strange vole of either sex into a cage with a resident pair, and the introduced voles subsequently show significant adrenocortical and splenic hypertrophy and thymic involution. These changes in organ weights were attributed to the fighting itself, although they may have resulted from competitive behavioural factors *per se* rather than fighting. This question was partially answered by showing that adrenal weight was inversely related to social rank in male house mice. Mice in a group rank themselves in a social hierarchy of dominance—subordinance relationships. The mean adrenal weight was least in the most dominant and greatest in the most subordinate mice. The experimental design was such that food or the ability to feed was not a problem. Comparable results were obtained with Norway rats. Subordinate males showed a marked decrease in adrenocortical sudanophilia when subjected to severe fighting for short periods, but normal sudanophilia and adrenal hypertrophy followed exposure to some weeks of less severe experience. The adrenals of the dominant individuals showed little change, although they fought as much or more than the subordinate rats. There apparently is an inverse relationship between social rank and adrenocortical activity in male rats and mice which is unrelated to fighting *per se*.

In our experiments severe fighting usually followed placing mice together for the first time, but usually ceased within an hour or two. The backs of subordinate animals were scarred in some populations, but more often scarring was rare or absent. These and other observations support the belief that fighting *per se* had little influence on the degree of adrenal enlargement and reproductive suppression.

Thymic involution in voles and changes in adrenal sudanophilia in rats indicate increased adrenocortical function with adrenal hypertrophy. Bullough (1952) showed that "crowding" produced marked cortical and medullary hypertrophy of the adrenals of mice associated with a 60 per cent suppression of epidermal mitotic activity which was considered indicative of increased adrenal secretory activity. A consideration of adrenal geometry will show that medullary hypertrophy would not contribute significantly to increases in weight compared to the cortex.

Other experiments have shown that increased population density depresses reproduction. Reproductive performance of albino mice declined with increasing population size and was attributed to behavioural factors. Retzlaff (1938) noted that the socially dominant female mice in each population reproduced best. Estrus cycles occur earlier, are more frequent, and last longer in segregated than in grouped mice. The latter experiments show that male mice are not required to elicit responses to increased density.

These experiments show that social competition can increase adrenocortical activity and decrease reproductive performance. However, if social competition is basically responsible for regulating population growth, there must be quantitative relationship between density and the magnitude of responses to it.

When previously isolated male mice were placed together in groups of varying sizes, adrenal weight increased approximately in proportion to the logarithm of the population size. Food and water were provided in sufficient quantity and from enough sources to constitute unlimited availability for all practical purposes. Adrenal weight in wild-strain house mice similarly increased with increasing population size but at double the rate seen in albino mice, presumably because of the greater reactivity, excitability, and aggressiveness of these mice. Hyperplasia and cellular hypertrophy of the zona fasciculata accounted for the adrenal hypertrophy. In the largest populations used for each strain of

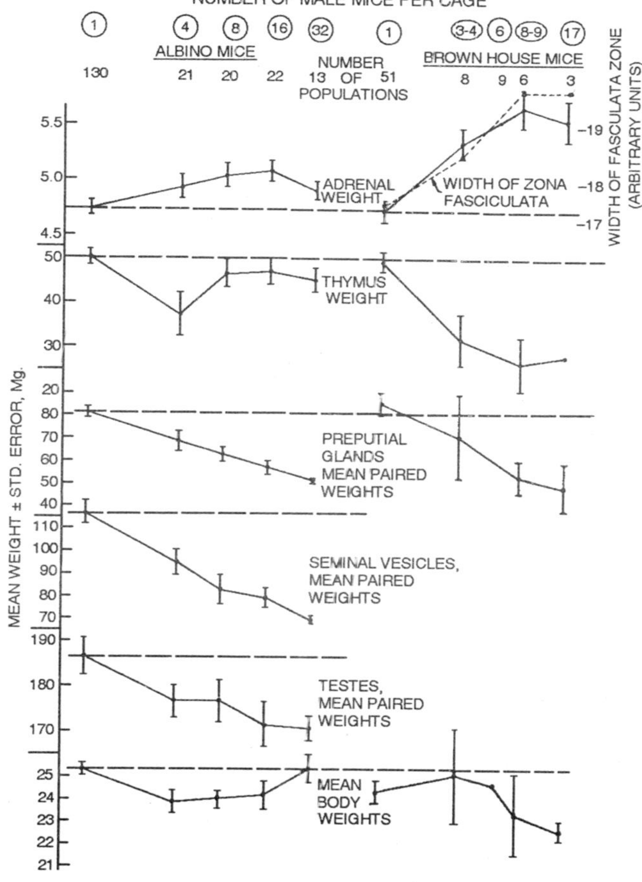

Fig. 12.1. Mean weights of the adrenals, thymus, and reproductive organs of male mice, albino on the left and wild strain on the right, from populations of fixed size plotted against population size on a logarithmic scale.

mouse (16 for wild-strain and 32 for albino house mice) the mean adrenal weight decreased from that found in populations of half the maximum sizes. A marked depletion of lipid in the zona fasciculata accounted for the decrease in weight although the hyperplasia was greatest in the largest populations. Thymus and preputial glands and seminal vesicles decreased progressively in weight with increasing population size. The weights of the preputial glands and seminal vesicles decreased almost linearly with increases in the logarithm of the population size. There was a suggestion of a loss in body weight with increasing population. The progressive thymic involution with increasing density was attributed to increased adrenocortical activity and not to androgens.

The progressive changes in organ weights were not related to increasingly limited space per mouse with increased population size. Adrenal weight reflected changes in population size to the same degree in cages with 42 times the floor area as those used in the preceding experiments. There must be a maximum amount of space that would still permit the mice to interact, but it is not within the limits of these experiments.

Decreased resistance to disease appears to be an important factor affecting mortality rates, and may have been responsible for increasing mortality high populations of voles and rabbits. Adrenocortical anti-inflammatory activity may be sufficient at higher population densities to decrease host resistance to infections. Cortisone and hydrocortisone are known to diminish resistance to and intensify the severity of infections. However, this thesis had to be tested. The amount of granulation tissue induced by implanting cotton pellets subcutaneously was significantly reduced (approximately 20%) by keeping mice in groups. The effects of crowding in suppressing the inflammatory response to cotton paralleled the effects of injected cortisone or hydrocortisone. Since cortisone is also known to decrease resistance to trichinosis, Davis and Read (1958) fed a fixed dose of embryonated *Trichinella* larvae to mice, some of which were then placed in groups and the remainder kept isolated. Three weeks later worms were recovered from the gastrointestinal tracts of 25 per cent of the isolated and 100 per cent of the grouped mice. The maximum number of worms recovered from an isolated mouse was less than the minimum number recovered from a grouped mouse. In another experiment the mice were segregated long enough following grouping to allow the *Trichinella* larvae to encyst. The infection rates of the isolated and grouped mice were approximately equal, but twice as many encysted larvae were recovered from the grouped mice and there was no overlap in the number of larvae from any one of the isolated with any of the grouped mice. The ability of *Trichinella* larvae to penetrate the intestinal wall is inversely related to the magnitude of the inflammatory response. Therefore the greater invasiveness of the larvae in the grouped mice suggests a suppression of the intestinal inflammatory response by endogenous factors. These experiments indicate that grouping mice depresses inflammation significantly, possibly by increasing the production of glucocorticoids. There appears to be some fact in the thesis that resistance decreases with increased population size.

Reproductive activity of both male and female mice declined with increasing population size in the preceding experiments, but the specific effects on the female mice were not determined. In a later experiment ten male and ten female mature mice were placed in a cage for six weeks and the data on female reproduction compared with those from littermate segregated pairs. All females in both groups became pregnant but seven of the ten crowded females lost their litters *in utero*, while the other three bore significantly smaller litters than the isolated females (7.67 vs. 9.00 pups per litter). The number of implantations was 53 per cent greater in the isolated mice. The time elapsing until the birth of the first litter was significantly longer in the grouped mice

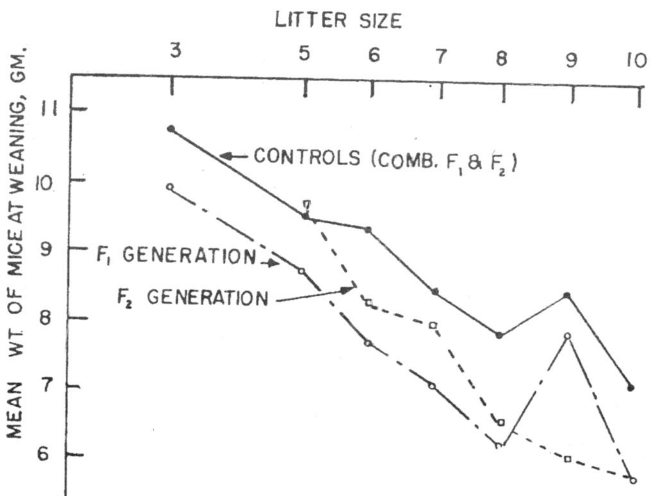

Fig. 12.2. Mean weights at weaning of progeny (F_1) of mice nursed by previously crowded mothers, their progeny in turn (F_2), and those of segregated controls plotted against litter size.

than in the controls. Grouping decreased the number of implanted embryos and increased intrauterine mortality. Fewer viable ova, failure to implant properly or both may have decreased the number of implantations.

Experiments with voles led Chitty to suggest that lactation was suppressed by crowding (1955). Similar experiments with albino mice showed that placing 20 males and 20 females in a cage together for six weeks resulted in (1) a marked depression of reproduction and (2) a suppression of lactation as indicated by diminished growth of young nurtured by previously grouped females. Diminished growth of the nestling young, apparently resulting from the original depression in lactation, persisted into the next generation, and was most marked in the larger litters. The effect on the growth of progeny in the first generation apparently was due directly to deficient nutrition during lactation, but the persistence of this effect into the next generation cannot be explained. In summary, increasing population density adversely affects the entire reproductive process of female mice, paralleling its effects on the secondary sex organs of males.

Experiments with Freely Growing Confined Populations

Under the artificial conditions of the preceding experiments a widespread physiological response to population density was reflected in a progressive adrenal hypertrophy, thymic involution, and reproductive retrogression with increasing population density. Increased adrenocortical activity was indicated by involution of the thymus and suppression of inflammation, while decreased androgenic activity was suggested by the weights of the male reproductive organs. The results were comparable for voles, rats, and two strains of house mice. However, one cannot conclude from these results that similar changes occur in growing populations. Nevertheless, mice in freely growing populations do respond to population density in the same way as those from artificial groups.

Populations were established by introducing a few pairs of house mice into a large cage with food, water, nest space, and nesting material well distributed and in excess of usage. Each population was allowed to grow without interference except for censuses and was sacrificed when growth had essentially ceased. Brown (1953), Strecker and Emlen (1953), and Southwick (1955a, b) had shown that the growth of confined populations of house mice was self-limited even with food and other necessities *ad lib*. Six of our populations were allowed to reach asymptotic or maximal values before being sacrificed, and two were sacrificed at one half of the estimated maximum size.

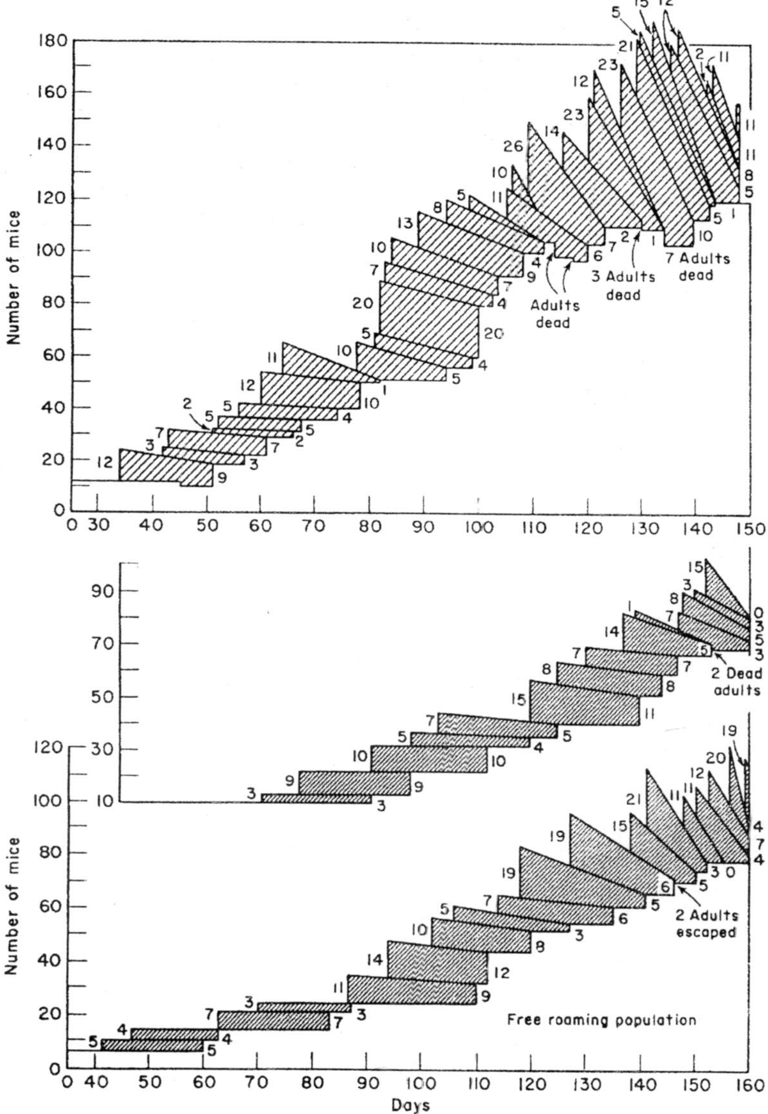

Fig. 12.3. Three experimental confined populations of house mice which were allowed to grow until they approached asymptotic values.

Birthrates and infant survival rates (to weaning) decreased approximately linearly as the logarithm of the population size increased. Infant survival approached zero as each population approached its upper limit. Those still surviving were weaned too soon and were stunted

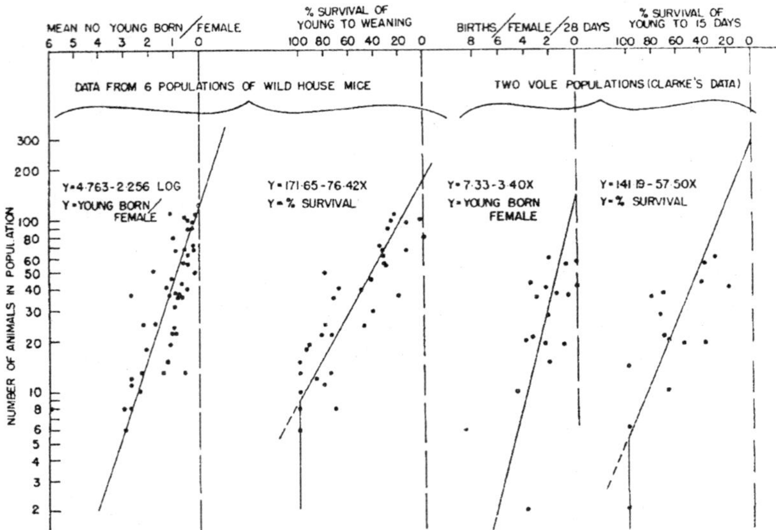

Fig. 12.4. Births per female and survival rate of the litters as percentages of those born for each increment of population for house mice from freely growing confined populations plotted against the logarithms of the population sizes.

and appeared to be in very poor condition. The stomachs of nestlings contained very little or no milk, even though nest destruction and interference with nursing by other mice were not observed. It was inferred that the high mortality rate of the young resulted from suppressed lactation. The declines in birthrate with increasing density indicated a progressive suppression of other reproductive processes. The prevalence of pregnancy was reduced 18 per cent, the number of embryos per pregnancy reduced 13 per cent, and the number of resorbing embryos per pregnancy increased 58 per cent in mature females compared to the control mice. The overall *ante-partum* losses in the number of viable embryos was at least 22 per cent. Southwick (1955a, b) also found a decline in fecundity associated with rising density in four of six freely growing confined populations of house mice.

Changes in the adrenal glands of both sexes and reproductive organs of the male mice were similar to those seen in the earlier experiments. Adrenal weight increased approximately 25 per cent in male and 14 per cent in female mice from high populations and about half these amounts in the intermediate populations. The increase in adrenal weight was largely due to cellular hypertrophy and hyperplasia of the zona fasciculata, although delayed involution of the X-zone

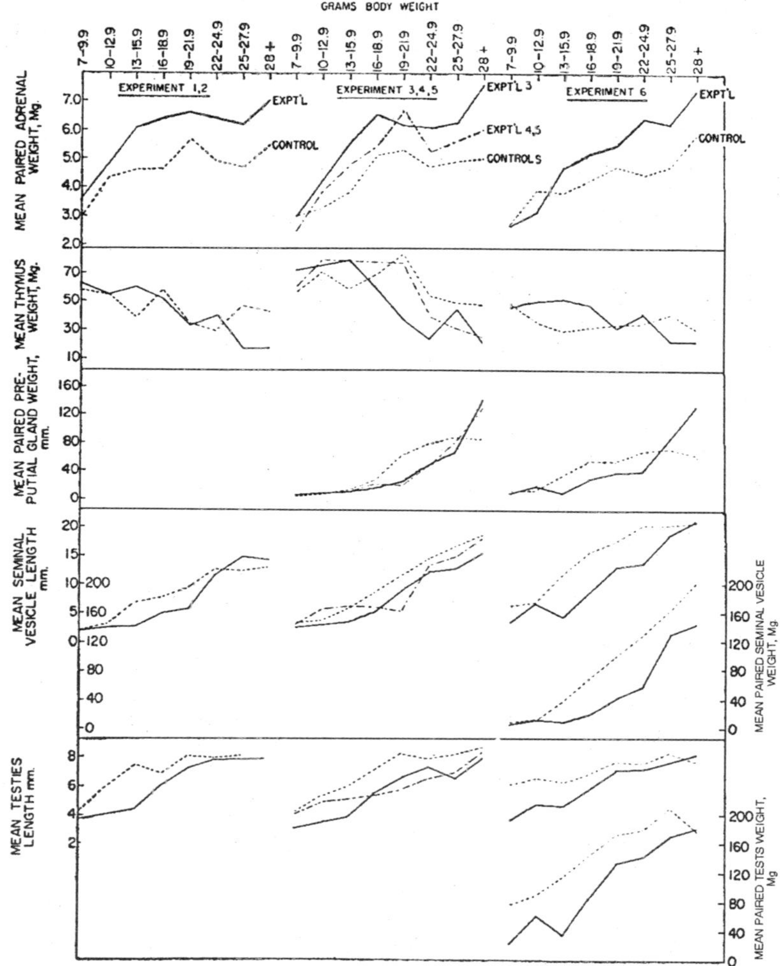

Fig. 12.5. Plots of the mean values for the sizes of the adrenal, thymus, and preputial glands and the testes and seminal vesicles of male mice from freely growing population of high (solid lines) and intermediate density (interrupted lines) and their segregated controls (dotted lines).

contributed in young mice. Delayed involution of the X-zone in male mice suggests that androgen production, therefore the onset of puberty, was significantly later in male mice from high populations and somewhat less so in those from the intermediate populations. The weights of the reproductive organs of male mice were significantly reduced except for those from mice too young to have begun sexual development, or from mice in the greatest body-weight group. The onset of

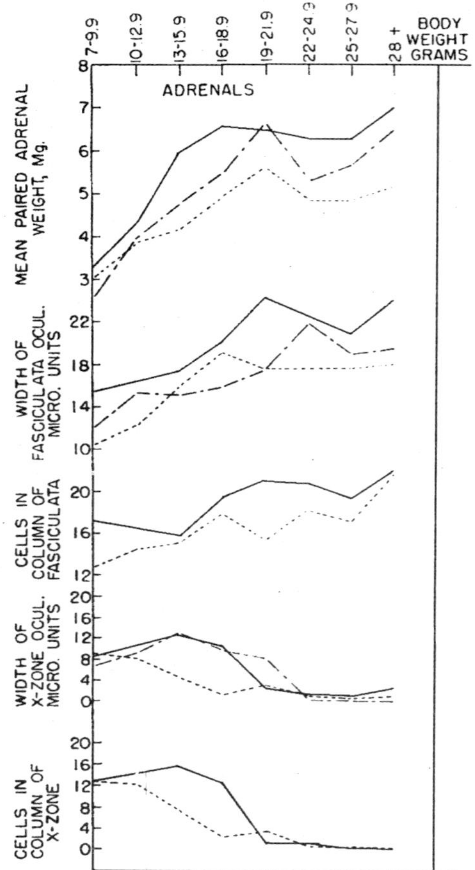

Fig. 12.6. Changes in the widths and numbers of cells comprising straight cords of the zona fasciculata and X-zone of male mice from freely growing populations.

spermatogenesis also was delayed with respect to body weight. The magnitude of comparable changes was approximately half as great in the mice from intermediate populations. Mice from freely growing populations exhibited the same responses to increased population density as mice in populations of fixed size. Adrenocortical activity increased and reproductive function decreased progressively in both sexes with increasing population size. These density-dependent endocrine adaptive responses to population density apparently were triggered by social competition since other factors, such as deficiencies of food, water, and nest space, which might have done so (either directly or indirectly by increasing competition) were provided in excess at all times.

A comparable experiment with meadow voles, *Microtus pennsylvanicus*, showed that these animals respond like house mice to increased population density, although the behaviour of these two species was quite different. House mice were essentially nonterritorial and huddled in large groups at higher densities, while the voles were distinctly territorial, occupying and defending a particular area of the cage against intruders. Even so, the voles had a social hierarchy within this arrangement. In this population the adrenals of mature male voles averaged 39.6 per cent heavier and those of mature females 36.6 per cent heavier than their respective segregated controls. There were no significant differences in the weights of the reproductive organs or thymus glands although the number of animals was too small to show significance considering the large variances in the weights of these organs. Clarke (1955), in similar experiments with two populations of voles, found that the birth and infant survival rates both declined as the populations increased in size. Louch (1956), also using confined freely growing populations of voles, found significantly lower eosinophile counts in animals from high density compared to those from low-density populations. He also found a significant negative correlation between fecundity and density for one of his three populations and a negative correlation approaching significance in the second. His third population showed a good positive correlation between adult mortality and density, whereas there was no correlation between these factors in the first population which had a significant fecundity-density relationship. Fighting was related to density in only the third population in which there was an appreciable adult mortality. A significant but smaller mortality of adults was seen in the second population. These results suggest that reproduction was significantly depressed at high densities in some way inversely related to mortality. A lessening of the degree of reproductive suppression may serve to compensate for appreciable adult mortality. We observed improved survival of young in a population of house mice which had a significant mortality of adults. The combined mortality of adults and young was the same as litter mortality alone in populations without adult mortality. Voles, like house mice, respond to increased population densities with increased adrenocortical activity, suppression of reproduction and increased infant mortality.

Experiments with Natural Populations

In order to show that internal feedback systems are basic regulators of population growth it is necessary to demonstrate the existence of mechanisms in natural populations of a variety of species. Data are available for a few natural populations of a limited number of species.

Adrenal weights were collected every four weeks for two years from a rural population of Norway rats. An index of population size was determined at each sampling. Adrenal weight for both sexes correlated significantly with the population indices for the two-year period. The weights of the pituitary and adrenal glands also were correlated significantly, but the correlation between pituitary weight and population index was less significant. A similar relationship between adrenal weight and population density was demonstrated by 49 samples from 21 populations of urban Norway rats. These populations were categorized as low stationary, low increasing, high increasing, high stationary, or decreasing, progressing in relative density from low increasing through the various stages to decreasing in the order. Adrenal weights increased as the relative density increased. The adrenals of rats from decreasing populations were 19 per cent heavier than those from low-increasing populations. The uncertain status of low-stationary populations has been discussed elsewhere. The adrenals from both sexes reacted equally to changes in density.

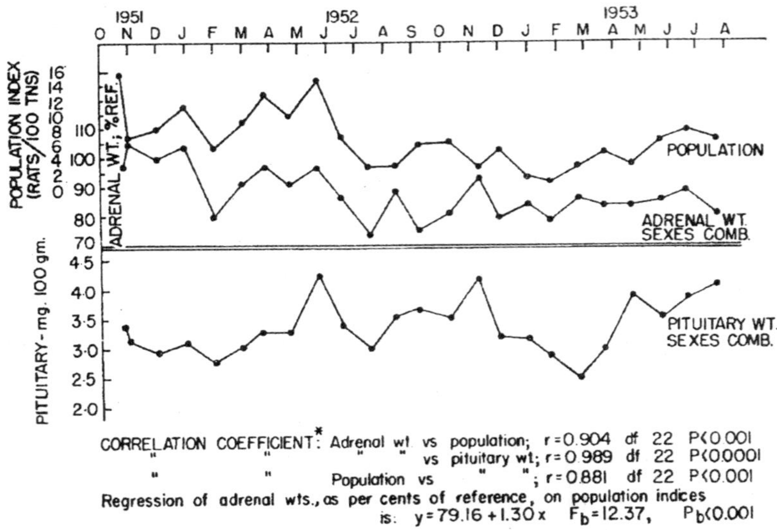

Fig. 12.7. Changes in population density, adrenal weight, and pituitary weight for a rural population of Norway rats which has followed by sampling at intervals of four weeks for a period of two years.

Reproduction is suppressed in rats from high populations. During the main spring breeding season, the prevalence of pregnancy was

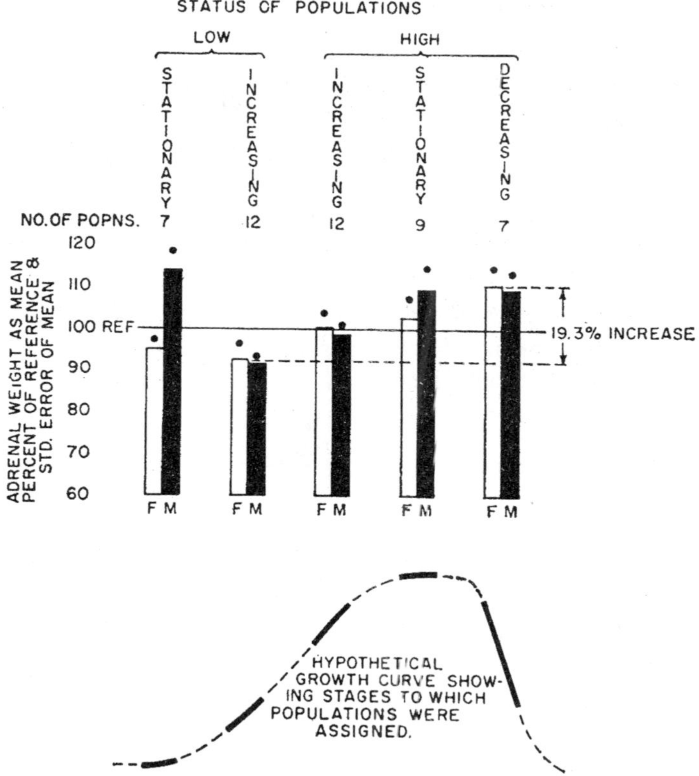

Fig. 12.8. Mean adrenal weights and their standard errors as percentage of reference values plotted against the density status of the populations from which the samples were taken for urban populations of Norway rats.

41.6 per cent in rats from increasing populations and 14.4 per cent in stationary populations. Mean litter size did not change with density. Increased density appears to affect reproduction in rats by reducing fertility and increasing litter mortality. The offspring of the subordinate rats have much less chance of surviving until weaning than those of dominant rats in a population.

If the adrenals respond to population density, then an artificial reduction in density should reduce the mean adrenal weight. Three populations of rats were reduced an average of 38 per cent from initially high levels and were maintained near this reduced level for a period of four months by trapping. The mean adrenal weight for both sexes declined 32 per cent following the reduction, and averaged 30 per cent below the original adrenal weight for the four-month period.

Further experiments with Norway rats implicated social competition in the regulation of population growth, especially in producing declines in population size. Introducing 22 per cent alien rats of either sex into high stationary populations produced an average decline of 40 per cent. If 20 per cent of the rats in rapidly increasing populations were removed and replaced by aliens of either sex, the growth of the populations promptly ceased. Low stationary populations remained the same when small numbers of alien rats were substituted for residents. Adrenal weights reflected the changes in population density. These experiments indicate that introducing alien rats into a population profoundly affects a population depending on its status at the time of the introduction.

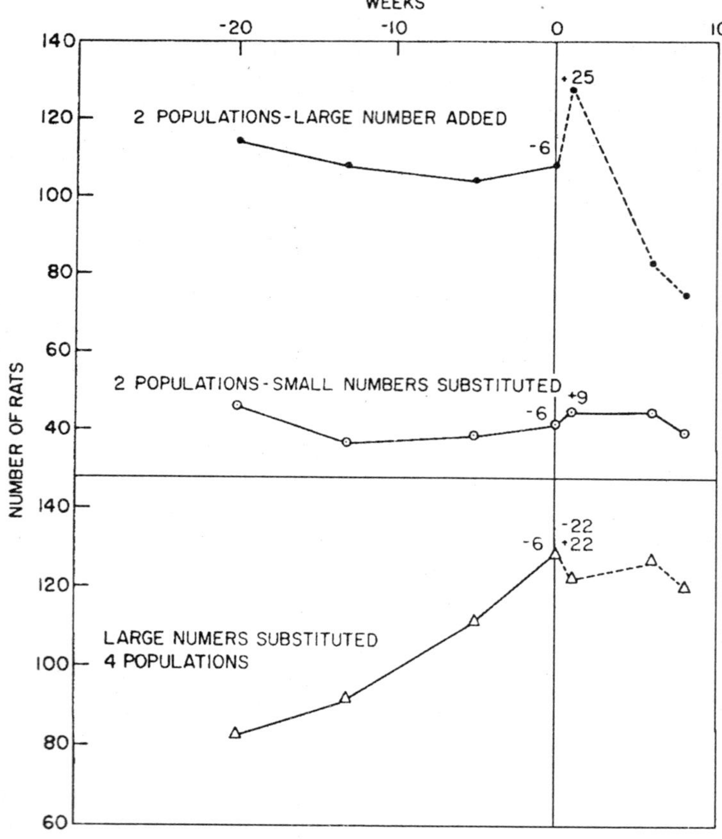

Fig. 12.9. Effect of introducing alien rats into populations in three different categories of density.

Several investigations on natural populations of several species of voles and other small mammals have given fairly dependable results. Adams, Bell, and Moore followed a population of *Microtus montanus* for five years beginning in 1951 with a period of high-population density. An index of relative population density was determined for each four-night sampling period. The mean adrenal weight, as a mean percentage of reference values, was determined from a portion of the female voles in each sample beginning early in 1952. There were too few male voles in most samples for valid comparisons. The prevalence of pregnancy was expressed as a per cent of the adult females which were pregnant. Significant changes in adrenal weight were related to changes in the relative population density. The population increased each spring and summer with maximum population densities in early fall, and the maximum adrenal weight each year occurred in late summer or early fall with a suggestion in 1952 and 1953 that it coincided with the beginning of decline in the population. The early phase of a decline in population size represents maximum relative density. The maximum prevalence of pregnancy each year preceded the maximum population levels. The numbers of females available to determine pregnancy rates were too small in the spring of 1953 to have any significance. Population densities and adrenal weights were greatest in

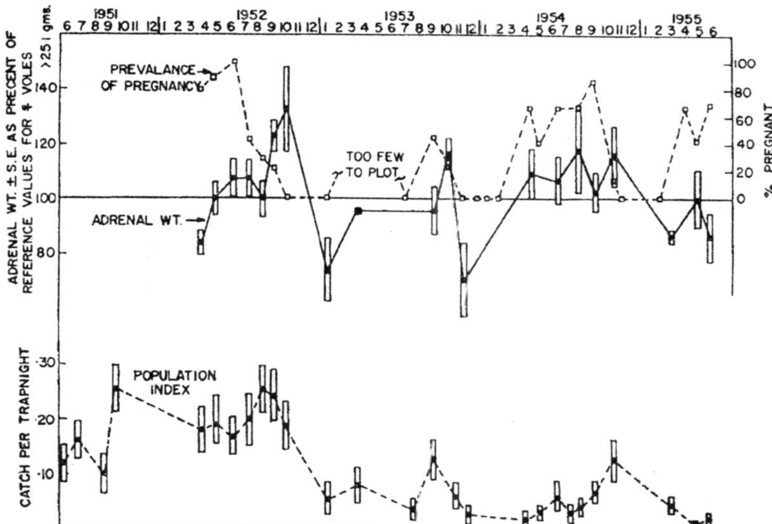

Fig. 12.10. Population density, adrenal weights and prevalence of pregnancy plotted for the females from a natural population of voles (Microtus montanus) which was followed by sampling for a period of more than four years.

September and October of 1952 when the prevalence of pregnancy was falling rapidly, whereas it remained high during the same period for the following two years. Two conclusions which may be drawn are that (1) changes in adrenal weight are closely related to population density, and (2) adrenal weight reflects the annual changes in population density as well as those occurring over a period of years. The increase in adrenal weight with the beginning of an autumnal decline in density may reflect increased competition resulting from the annual decline in the abundance of available food supplies; but it does not reflect environmental changes directly or the adrenal weight would not follow the decreases in the population to minimal levels during the late fall and winter of every year. One of the objectives of this study was to "detect and study any diseases that appeared and which might affect the course of the (population) cycle." Dr. J. F. Bell was unable to isolate any pathogens from the few animals with lesions suggestive of infectious disease. Most of the animals appeared normal at necropsy. The marked decline in the vole population from a peak in the late summer of 1952 appears to have been due to physiologic alterations resulting from very high population densities. Chitty (1954) reached a comparable conclusion in a study of a population of voles. In two natural populations of voles (*Microtus pennsylvanicus*) Louch (1958) found that eosinophile counts were low during and immediately after a period of high density and then rose about fivefold and remained high in both populations for the duration of the study. In one population male adrenal weights were maximum during the period of high density and declined an average of 37 per cent for the remainder of the study when the population was low. In the second population the mean adrenal weight in the fall of the first year was less than for the high population during the same period, but it was significantly greater than for the remaining period in the same population. The second population may have experienced a decline from much higher levels before the study began.

Hoffman (1958) found an inverse relationship between population density and natality, in the voles *Microtus californicus* and *Microtus montanus*. Ovulation rate and litter size were low during a decline in density of the vole populations. Post-natal and weanling mortality was high during peak densities and subsequent declines, but was low during the phase of population increase. The prevalence of pregnancy declined as density increased in the populations of voles (*Microtus pennsylvanicus*) studied by Hamilton (1937).

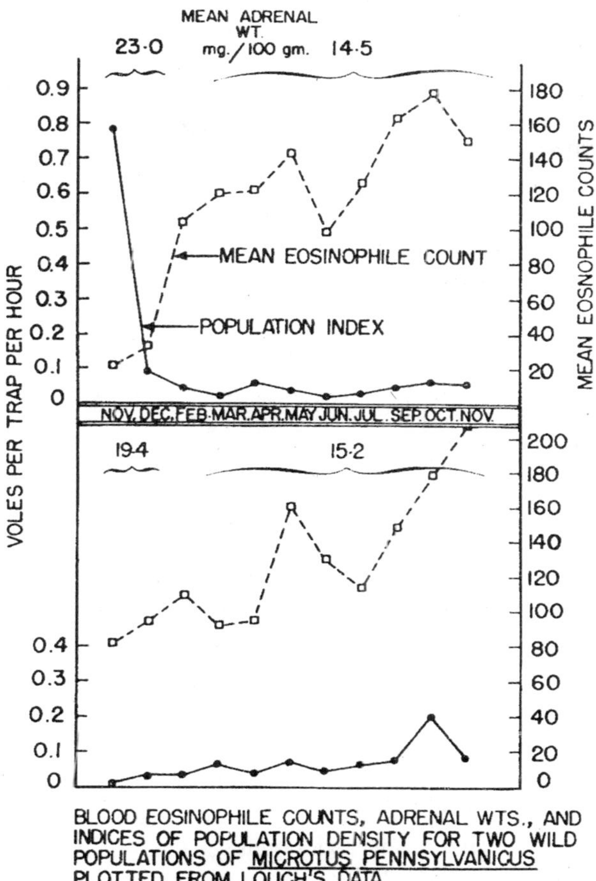

Fig. 12.11. Data for two natural populations of voles (Microtus pennsylvanicus) reported from Louch.

Kalela (1957) studied populations of the vole *Clethrionomys rufocanus* for three years. These voles achieved moderate densities in the first summer, more than doubled their numbers in the following summer; in one local area of study the latter densities were doubled. The populations collapsed in the fall after reaching peak densities and were at very low densities by the third summer. There were no food shortages in peak densities or when the population declined abruptly during the following winter. Food utilization was very slight compared to that available. There were no significant changes in litter size associated with population density. During peak densities juvenile voles, especially males, failed to reach sexual maturity, usual by late summer.

Juvenile females also failed to reach sexual maturity in the area of extreme density. The prevalence of pregnancy in fecund females in late summer was lower in high than in populations of low densities. In short, breeding ended earlier during the period of high density. Muskrats (*Ondatra zibethica*) also stop breeding earlier in periods of high density. Kalela concluded (1) that the fecundity rate of juvenile voles is density-dependent, (2) that variability in fecundity rate is the most potent means of regulating reproduction in a species, and (3) that the regulation of fecundity rate is not due, directly or indirectly, to shortage of summer food but to intraspecific intolerance with increasing numbers.

Kalela's results agree with those of other studies we have discussed. The attainment of sexual maturity of house mice is delayed at high population densities, but intrauterine mortality is also appreciable. These differences between species are probably not as great as they appear. Failure to reach maturity may be a response to a more severe stimulus than that represented by *in utero* losses. In one population of house mice, which was permitted to survive after reaching maximum density, the birthrate fell to zero and remained there.

In a less detailed study of the effects of population density on white-footed mice, *Peromyscus leucopus*, it was found that adrenal weight dropped 58 per cent ($P < 0.001$) while the index of relative population density declined 68 per cent from one July to the next.

None of these experiments with natural populations of voles or white-footed mice is by itself proof that the same endocrine responses to increased density occur under natural conditions that occur in experimental populations, but together they offer convincing evidence that such is the case.

INDEX

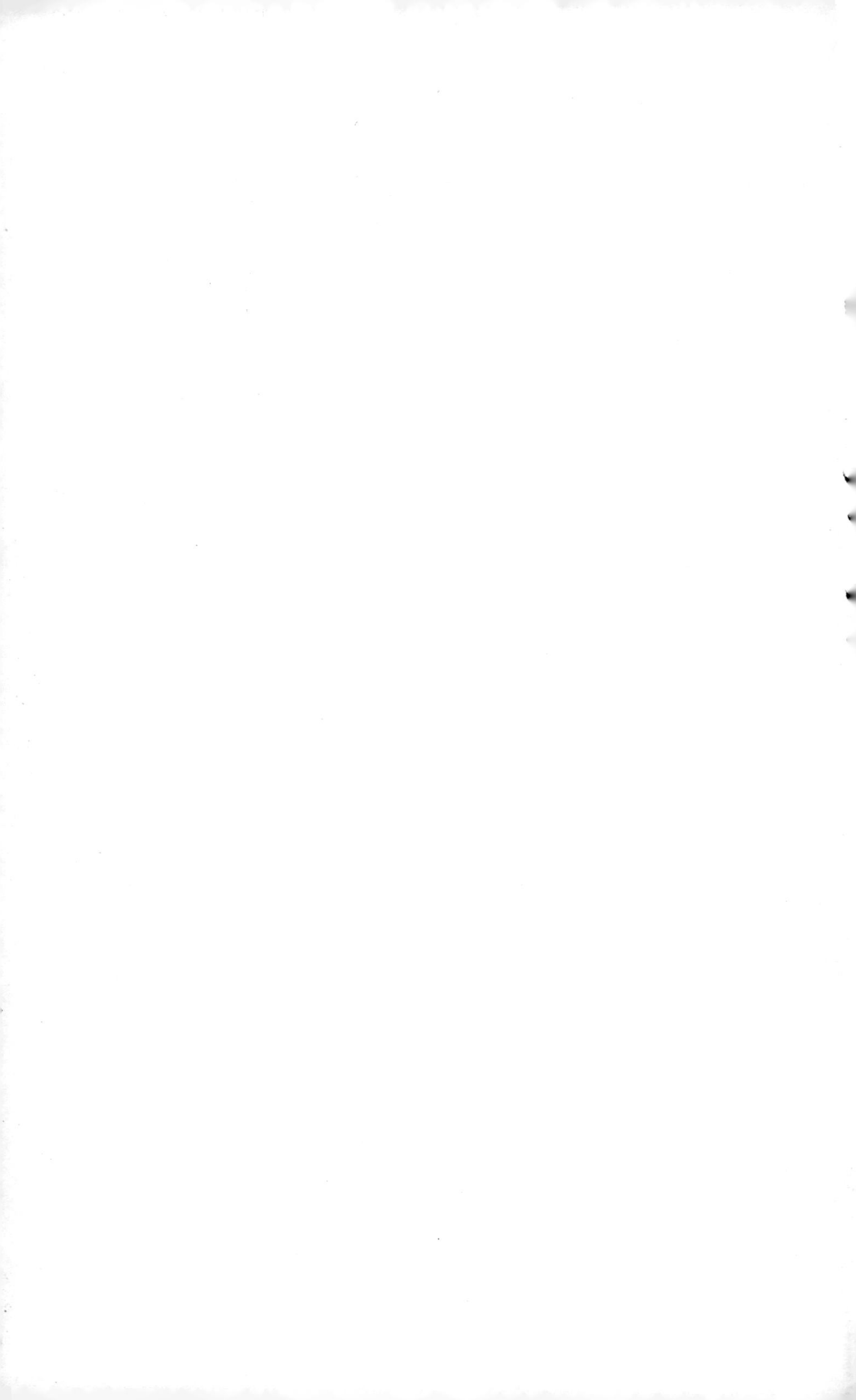